Yesterday's Farm Tools & Equipment

Featuring Collections of the Landis Valley Museum

Michael B. Emery and Irwin Richman

Schiffer Publishing Ltd®

4880 Lower Valley Road, Atglen, Pennsylvania 19310

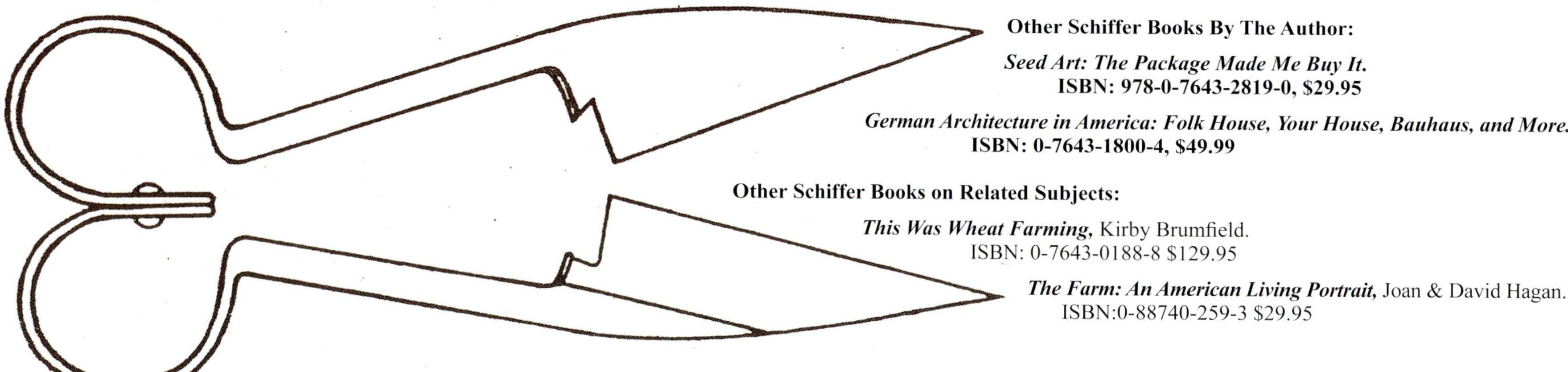

Other Schiffer Books By The Author:

Seed Art: The Package Made Me Buy It.
ISBN: 978-0-7643-2819-0, $29.95

German Architecture in America: Folk House, Your House, Bauhaus, and More.
ISBN: 0-7643-1800-4, $49.99

Other Schiffer Books on Related Subjects:

This Was Wheat Farming, Kirby Brumfield.
ISBN: 0-7643-0188-8 $129.95

The Farm: An American Living Portrait, Joan & David Hagan.
ISBN:0-88740-259-3 $29.95

Library of Congress Control Number: 2010930822

Items are from the Landis Valley Collection, Lancaster, Pennsylvania, unless otherwise credited.

Designed by Stephanie Daugherty
Type set in Aldine721 BT/P22 Cezanne Regular/Times New Roman/ Times New Roman
ISBN: 978-0-7643-3603-4
Printed in China

Schiffer Books are available at special discounts for bulk purchases for sales promotions or premiums. Special editions, including personalized covers, corporate imprints, and excerpts can be created in large quantities for special needs. For more information contact the publisher:

Schiffer Publishing Ltd.
4880 Lower Valley Road
Atglen, PA 19310
Phone: (610) 593-1777; Fax: (610) 593-2002
E-mail: Info@schifferbooks.com

For the largest selection of fine reference books on this and related subjects, please visit our web site at:

www.schifferbooks.com

We are always looking for people to write books on new and related subjects. If you have an idea for a book please contact us at the above address.

This book may be purchased from the publisher. Include $5.00 for shipping. Please try your bookstore first. You may write for a free catalog.

In Europe, Schiffer books are distributed by
Bushwood Books
6 Marksbury Ave.
Kew Gardens
Surrey TW9 4JF England
Phone: 44 (0) 20 8392 8585; Fax: 44 (0) 20 8392 9876
E-mail: info@bushwoodbooks.co.uk
Website: www.bushwoodbooks.co.uk

Contents

Dedication

Mike dedicates this book to his parents,

Logan and Ruth Ann Emery;

to his wife,

Alia,

who has been supportive (or at least tolerant) of

his historical interests and his collecting habits;

and their son,

Joshua.

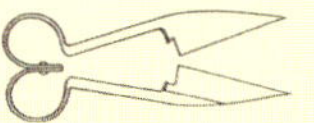

Irwin dedicates his work to his wife,

Sue,

who countenances and usually encourages his interests;

their sons, Dr. Alexander Richman, a mathematician,

and Dr. Joshua Richman, a medical researcher;

their grandchildren;

and honorary son,

Oscar D. Beisert, Jr.,

an architectural historian.

Acknowledgments

Two very different people from different backgrounds came together to write this book. One is in his seventies and the other in his thirties. Special thanks go to James Lewars, Site Administrator of the Daniel Boone Homestead, who is a font of information about rural architecture and meadow irrigation; and Lester Breiniger, Jr., potter and retired educator, whose knowledge of Pennsylvania German folkways is encyclopedic. The librarians at the Pennsylvania State University at Harrisburg have always helped us find occasional obscure information.

Of special importance are our colleagues at the Landis Valley Museum: Stephen Miller, past site Administrator of Landis Valley (he is now Director of the Bureau of Historic Sites and Museums of the Pennsylvania Historical and Museum Commission), helped launch this project. His successor, Russell Swody, has given us his complete cooperation, Bruce Bomberger, Curator of the Landis Valley Museum, is expert on the agricultural and technology collections of the museum. Joe Schott, our Farm and Garden Manager, is an authority on historical agriculture. Craig Benner has done all of our photographic work. Our other colleagues have always helped when they could. Thank you, Donna Horst, Nicole Wagner, Jen Glass, Tim Essig, Will Morrow, Joyce Perkinson, Cindy Reedy, and Trish Frey. Dr. Russell Eaton, a retired physicist, has been cataloging Landis Valley's postcard collection and has brought many items to our attention.

Mike is computer literate and Irwin is not. Special thanks go to Irwin's wife, Dr. M. Susan Richman, retired mathematician and University dean, for turning Irwin's scrawl into usable text and her editing help.

Chapter 1

Farming and Tools

Farming is the basis of man's progress towards civilization. The ability to provide a stable food supply marks the beginnings of the transformation of mankind from bands of hunter-nomads into a settled populous. Farming, undeniably a stabilizing influence, had also become a tradition-bound enterprise. Many of its techniques and tools became ritualized and unchanged for centuries and even eons. It is widely held that a farmer in medieval England would have been perfectly at home on a farm in eighteenth or early nineteenth-century America. Even today some farm implements remain recognizable, in form, from the dawn of agrarian civilization. Agricultural reformers on both sides of the Atlantic Ocean have, at times, been frustrated by this conservatism. English gentleman, Sir John Sinclair noted in his *Code of Agriculture,* published in 1817, that:

> The introduction of new implements into a district is often a matter of the greatest difficulty, owing to the ignorance, the prejudices and the obstinacy of farm servants and labourers. Many farmers therefore, very absurdly retain their old implements, though convinced of their inferiority, rather than sour the temper of their labourers by attempting to introduce new ones.

The ultimate triumph of industrialism in 19th century America led to a boundless enthusiasm for change among progressive farmers. George A. Martin, editor of *Farm Appliances: A Practical Manual*, published in 1887, joyously proclaimed:

> Inventive talent has completely revolutionized the processes of farming. The work which required the labor of many, under primitive methods, is now better done by one person with the aid of improved appliances. To explain and illustrate some of the most practical and easily made appliances is the object of this volume. They are such as secure greater comfort to domestic animals, provide supplies of wholesome water, economize labor and assist in dispatching much of the important work on the farm.

The pre-industrial farm is characterized by this 1916 barnyard scene photographed in south-central Pennsylvania. Cows and chickens wander and a monumental hog basks in the sunlight. Note the hay wagon -- a rustic form that, no doubt, would have been familiar to a 17th century American farmer.
Collection: Mrs. & Mrs. Michael B. Emery

A country auction photographed by Landis Valley Museum founder Henry K. Landis (1865-1955) early in the twentieth century shows an assortment of horse-drawn farm implements and vehicles awaiting sale as well as a man-powered wheelbarrow. The auctioneer visible above the first carriage from the right commands the crowd's attention, including the young women in the open barn doors.

Nostalgia for the good old days revolves around an interest in pre-industrial farm life. In this carefully posed photograph a knickers-clad boy brings water to a plowman. It would have been a long walk for one cup. Where is his bucket or jug? As the farmer pauses at his plow, his team rests and the double tree, which helps coordinate the team, lies on the ground. *Collection: Mrs. & Mrs. Michael B. Emery*

A decade earlier, Isaac H. Hillborn, of Newtown, Bucks County, Pennsylvania, had published an article in an agricultural yearbook entitled *Machinery on Small Farms*, which begins with the question: "Has the introduction of labor-saving machinery been an advantage to the farmers of small areas?" His answer, an enthusiastic "Yes," is cast against a historical background:

> The introduction of labor-saving machinery has not been a thing of a day, or a year; but has been the work of ages, and in a measure has been the result of a necessity. In ancient times it was but needful to tramp the grain into the soil by the hoofs of the cattle, and later but to loosen the soil by means of a crooked stick fastened to the tail of the patient ox. Then, could all the corn needed be ground by pestle and mortar; but, in lapse of time, as population increased, more land must be tilled, more grain must be produced, and more bread must be made. To accomplish these results gave rise to the necessity of labor-saving machinery. Then was the rude plow fashioned, with curved mould of wood, and share of same material bound with

The farm tools of the mid-19th century would have been recognizable to a medieval farmer – and a farmer in pre-World War II America. On the frontispiece of a German language farmer's almanac published in 1852, a rising sun lights up a prosperous farm with neat rows of corn shocks. In the mid-ground a plowman works with a two-horse team and in the foreground the tools of the farmers' craft are displayed along with sheaves of wheat and an active beehive.

Farm implements are designed for work, but mankind is also a playful animal who can find humor and enjoyment in varied familiar venues. An exuberant mustached farmer frolics as an exclamation point on a mower while his dour colleague, eyes downcast, holds the reins of the team; three adventurous women balance on a harrow in a harvested cornfield; and a Sunday group, plays on and with a hay rake. *Mower & Sunday group. Collection: Mrs. & Mrs. Michael B. Emery.*

Farm machinery was a source of pride for farm families. On the Hoopes farm in Chester County, Pennsylvania, grandmother and granddaughter, Elizabeth Hoopes and Betty Lou Plank, pose before a massive turn of the 20th century Frick steam tractor engine. On the nearby Emery farm two brothers, Lewis and Ed, proudly show off a gasoline-powered Minneapolis tractor – of much smaller proportions. *Collection: Mr. and Mrs. Michael B. Emery.*

iron. This was a labor-saving implement of modern times, yet who would think of using it now?

As pestle and mortar were found insufficient, they were replaced by the water-mill, with all its modifications and improvements. As the sickle and scythe in the hands of the sturdy farmer were found unequal to the task allotted them, the mower, reaper, binder, and rake followed. The next akin to sickle and scythe—the flail—was hung upon some forgotten peg, and the thresher and cleaner took its place. The labor of seedtime and harvest has been made but a pastime. What once was the laborious task of days for all hands upon the homestead is now performed by the lads and lasses of the household, as though it were but recreation. What once required the muscular force of all the sturdy laborers that could be mustered is now accomplished by half-grown boys with the aid of a span of horses, which otherwise would have been idle. Not alone in the field and workshop has the influence of labor-saving machinery been felt. The burdens of the housewife have, in a measure, been lifted from her shoulders, though not in the same proportion as those of her husband. The washer, wringer, and the sewing-machine have each taken their part of the household labor, and performed it well.

Labor-saving devices and new machinery, while highly desirable, should be acquired selectively. "We do not assert," Hillborn wrote, "that every tiller of the soil should be supplied with all [possible]...implements, but some at least will be needed whether his lot be cast upon our eastern slopes or the broad western prairies."

Others perceived a downside to the embrace of machinery – in their view, industrial progress discouraged the hard work needed to cultivate crops requiring extensive handwork, and hence were seldom to be grown in America. Flax, the fiber plant source of linen, is one such product where "... we [Americans] are fast drifting into effeminacy [while] the hardy toilers of France, Germany, Holland and Ireland are supplying us ..." This was the view of Dr. J. F. Edge, of Chester County, in 1878. "My earliest and most pleasant memories," he wrote,

> as of most Pennsylvanians of my age, are closely associated with the spinning-wheels of our mothers, the distaff, the reel, and the breaker. The pulling, rotting, breaking, and hacking for the flax was a pastime for the boys of the farm, as the spinning, reeling, and twisting was of the girls. Each family was a community in itself. It is one of my proudest recollections that I

Many crops were or could be grown in varied areas of America. Hops is one of these. In the 19th century the area around Cooperstown, New York, became the hops growing capital of America. Thanks to the railroad and California's ideal hop-growing climate, by the end of the century virtually all hops came from the Golden State. Tall growing poles were unique to hop fields. *The Cottage Collection.*

In areas of America where fields were vast and flat, amazingly large teams were used to propel large combines as in this California scene, *circa* 1910.

assisted to produce the linen and woolen material of my youthful wardrobe. If the farmers' sons and daughters of to-day could make the same boast in the future, it were well for them, as evidence of a return to the primitive and frugal habits of long ago.

The good doctor was incensed, as a patriot, that America was importing $25,000,000 worth of flax a year, while "Three million women are employed in Russia in the spinning of flax by hand. "What," he asks rhetorically, "are our women doing?" Why not grow flax in Pennsylvania? His answer is, "… unthrift and want of better systems of industrial training for our unprofitable, because unskilled, hands."

Growing hemp for fiber production was largely abandoned in America by the third quarter of the 19th century, as was flax for making linen. Part of the process of loosening the fiber from the stalk involved using a brake to crush the stems. Recognized as an archaic farm tool early in the 20th century, this Hemp Brake, or *Hanf brecher* **(In Pennsylvania Dutch) was photographed by Henry K. Landis.**

Dr. Edge asked other questions about hemp production, as well as sericulture (raising silkworms and harvesting their cocoons). In both cases, in essence, he cites laziness for these trends. Like many others, including the Corn Law League in Great Britain, a generation earlier he was resisting the globalization of agriculture. The Corn Law people in the British Isles argued that all grain consumed there should be grown in the United Kingdom. This was doomed to failure because grain could be grown abroad more cheaply. Further, with expanded trade it made little sense for Americans to raise silk or hemp – while we were quickly becoming the world leader in the production of a more abundant vegetable fiber, cotton.

Cotton remains important in many southern states, but small quantities of the crop were raised in other areas, even in southern Lancaster County, Pennsylvania, into the early 19th century. While cotton picking would become highly mechanized, harvesting the crop by hand endured well into the 20th century. The field scene, complete with picking sacks, was mailed in 1969. *The Cottage Collection.*

As to the charge of laziness being raised as the reason these crops passed from local cultivation, it becomes questionable when compared to the labor entailed in growing tobacco, a crop raised on many a former flax field. Tobacco was and is one of the most labor-intensive major crops grown in America and, not incidentally, for over a hundred years, the most profitable.

The Pennsylvania Germans of south-central Pennsylvania are, paradoxically, among the most traditional and yet progressive of American farmers – and their love of the land is enduring.

"Among the Pennsylvania Germans," writes author Dennis Boyer,

> ...there is a nearly three century-old love affair with farming that is only now starting to dim due to land prices and urban sprawl. Undoubtedly this love affair goes back to Europe – to the cantons of Switzerland, to the fertile plains of the Pfalz and to the green meadows of Alsace – stemming from a time when family land ownership was beyond the grasp of ordinary citizens.
>
> But for the better part of two centuries, to be "Dutch" was to be a farmer. Even for town-dwellers in the region the real family home label was assigned to an uncle's or a grandfather's farm.

It was into this world that two ultimately foresighted and eccentric brothers, Henry Kinzer Landis (1865-1955) and George Diller Landis (1867-1954), were born. Both modern men of their time, each attended Lehigh University, and Henry Landis had a forty-year-long career in New York City, editing trade journals, mostly in the field of natural gas.

Both, but especially Henry, realized that the farming world they knew as children was disappearing and they consciously set out to collect the artifacts of a vanishing traditional culture. Among their eclectic accumulations was an extensive collection of farm tools and devices. This laid the foundation for the farm equipment collection of the Landis Valley Museum, near Lancaster, Pennsylvania, which is among the most comprehensive in America. It is, however, also regional and its examples relate to agriculture in the north-east with especial attention to those of south-central Pennsylvania. Accordingly, there are many artifacts relating to grain and tobacco culture, but none specific to growing cotton.

Drawing heavily upon Landis Valley Museum's vast farm tool collection, this book is not an encyclopedia, but rather an expanded vista of farming practices and farm devices commonly employed in America until the end of World War II, with especial emphasis on the period 1850-1918.

Probably photographed on Decoration Day, circa 1900, a group gathers on the porch of the General Store-Post Office of a small farm-serving town in Pennsylvania. To the right is one of the greatest engines of change in farming technology: the gasoline pump! *Collection: Mr. and Mrs. Michael B. Emery.*

Chapter 2

Pastures, Meadows and Haying

The sight of cows—brown, tan, white, black, or varicolored—in a pasture characterizes, for many, the quintessential image of farm. Artists of the Hudson River School, America's first native-born painting tradition, were devoted to portraying the American landscape, a landscape enhanced with cows ambling along streams and reposing in pastures. Cows are symbolic of peace, tranquility, and plenty. Cubist painter Arthur Dove's modernist "Cows in a Pasture" is a complex non-figural composition of swirling blues, browns and greens representing earth, pasture and sky and dynamic blacks and whites, evocative of cattle. In popular culture in 2009, the California cheese industry promotes advertisements, featuring anthropomorphic cows grazing in rich pasture land that proclaim, "Great cheese comes from happy cows. Happy cows come from California." This fantasy image is, of course, archaic, especially in California where most dairy cows are raised on what have been characterized as factory farms.

Another exemplar of farming life is haying and haystacks. Images of farm-folk haying and loading hay wagons are common themes of 19th century American landscape painting. A great painting by artist George Innes (1825-1894) called "Peace and Plenty," now at the National Gallery of Art, is a hayfield serenade or sonata. Haystacks have long attracted artists. French artist Claude Monet (1840-1926) is famed for his lyrical studies, as was American artist Martin Johnson Heade (1819-1901).

The reality of haying and stack building was hot, back-breaking, insect-infested labor. While artists have left us many visions of these rural activities, they are simply beautiful, devoid of sweat and aromas. Also missing is any sense of the dangers of haying. Sharp implements, heavy loads, and animal mishaps often made farming a dangerous occupation and hay fields were especially accident-intensive areas.

Pasture land and meadows are central to traditional American agriculture but they are not synonymous. Pasture is land devoted to grazing animals; meadows are areas devoted to growing grasses to harvest for winter use. 'Meadows' and 'hay fields' are often used interchangeably. One slight complication is that on many farms animals are allowed to graze on stubble after grass has been harvested and hay made.

Brooklyn-born artist Clinton Loveridge (1824-1902) often painted his Hudson River School inspired landscapes in the Albany, New York, vicinity. Farmers favored pasture land bordered by small streams where the cattle could self-water. Other cows graze in the background. Today in many areas farmers are compelled to fence along streams so that cattle waste doesn't foul water supplies. *The Cottage Collection.*

In their most primitive forms, both pastures and meadows depend on native vegetation, but man learned to improve both. Meadowlands were often mechanically enhanced. Rich, low-lying marshy land was often drained to provide stable land to farm upon and land that can support more desirable mixes of grasses, legumes and forbes. Dry meadowland near streams could be irrigated by construction of simple irrigation ditches or canals. Both kinds had to be kept clear, which often entailed much hard labor.

Historically the crops grown on pastures and meadows differed. Traditionally, meadows were as close to mono-cultures as technology allowed, while in pastures a wide variety of plants were encouraged. In the more scientific age of agriculture emerging after World War II, carefully controlled plantings were introduced into pastures and plant diversity declined. Miriam Rothschild, a British biologist and a scion of the famed banking family, in recent years led a movement to restore plant diversity to pasture land. Dr. Rothschild argued that it was better for the environment, the preservation of endangered species, and the enjoyment of animals that she believed were bored by the uniformity of modern pastures.

The most familiar tools associated with pastures and meadows are meadow or haying tools, especially scythes, cradles or rakes, hay forks and their like. Less familiar are meadow implements including clover seeders and especially specialized tools designed to be used in opening and cleaning drainage and irrigation tools.

Different meadow crops were popular at different times. These include mixed grasses, mono-cultures of specialized hay crops such as "Hungarian Grass," white clover, timothy grass, and alfalfa. Here three teams of mules propel mowers as they cut a large meadow in California.

Cattle stand in a pasture in front of the commodious, and still standing, 19th century Sheetz house in Berks County. A barbed-wire fence separates them from the rolling lawn surrounding this prosperous-sized farmhouse. *Collection: Mr. and Mrs. Michael B. Emery*.

In a carefully staged photograph an elegantly dressed woman uses a horse-drawn rake to show how easy this task is with modern machinery. To the left a boy uses a pitchfork to begin assembling the mown hay into stacks like those in the background. A stamp identifies the scene as being near Brunswick, Missouri, but this was a stock postcard sold as being representative of many areas. *Collection: Mr. and Mrs. Michael B. Emery*.

Farming near Allentown, Pa.

Once hay was cut, raked, and stacked, it had to be loaded on wagons for transport to barns or to massive storage-sized haystacks near the animal barns. Traditionally pitchforks were used for this difficult, hot task. Often ladders were used to allow hay wagons to be loaded to their maximum.

Manpower and horsepower are prominent in this New Hampshire haying scene. The huge hay rakes were used in many parts of the nation and are represented in the Landis Valley Collection. *Collection: Mr. and Mrs. Michael B. Emery*.

The "Haymow and Harness Room" of the Landis family barn, shown freshly filled in late summer, was photographed by Henry K. Landis, *circa* 1900.

Building haystacks for winter and early spring use on a prosperous farm was arduous, dangerous work often employing ladders. Passing the hay from level to level with pitch forks required great dexterity and endurance. *Collection: Mr. and Mrs. Michael B. Emery*.

In the American West, stacking hay (or alfalfa) was an extraordinary enterprise involving elaborate man- and animal-powered lifts. Stacks truly assumed monumental stature. Note the horse-drawn rake and mower posed in front of the medium-sized haystack. These circa 1910 post cards were printed in Los Angeles and Salt Lake City. *The Cottage Collection*.

A hay wagon newly arrived in a drive-through barn. Note the man at the left holding a pitchfork. He no doubt helped load the wagon and, after the horses were unhitched and tended to, would help unload the wagon's content into the hay mow. Single horse shafts rest against the barn. *Collection: Mr. and Mrs. Michael B. Emery.*

The horse and the gasoline engine meet. Horse-drawn equipment mowed and raked the hayfield. A motor-powered elevator loads the hay onto a heavily laden truck.

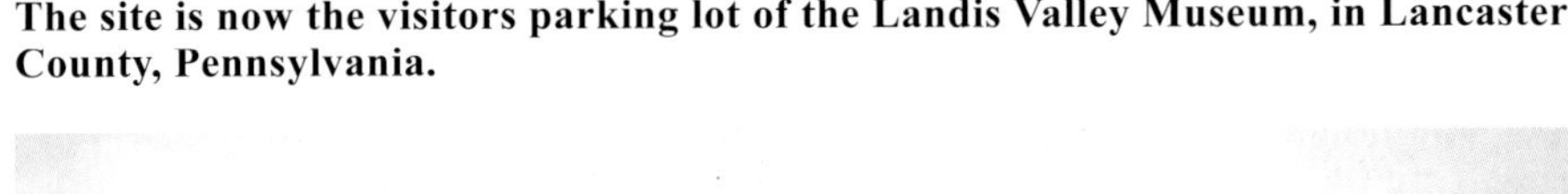

The site is now the visitors parking lot of the Landis Valley Museum, in Lancaster County, Pennsylvania.

Machine-baled hay waits to be moved into the barn for storage.

The triumph of the machine. Two men pose in front of a truck loaded with hay bales – the product of a mechanical baler. Freshly cut and raked hay awaiting processing is in the foreground.

Horse-drawn equipment harvests alfalfa hay to be baled on the Nebraska prairie. *Collection: Mr. and Mrs. Michael B. Emery*.

From the Landis Valley Collection

Stump Pullers. Used in clearing land. *Historic photograph by Henry K. Landis*

Meadow Hoes. Used to clear irrigation ditches.

Clover Seeder. Mike Emery serves as a proportion marker.

Scythe. The most common manual harvest tool for grasses.

Scythe blades.

Snaths. Name given to scythe handles.

Hay Mower.

Clover Cradles and detail. Specialized scythes designed for harvesting clover.

Clover Hullers. Prepare clover seed for planting by cracking tough hulls.

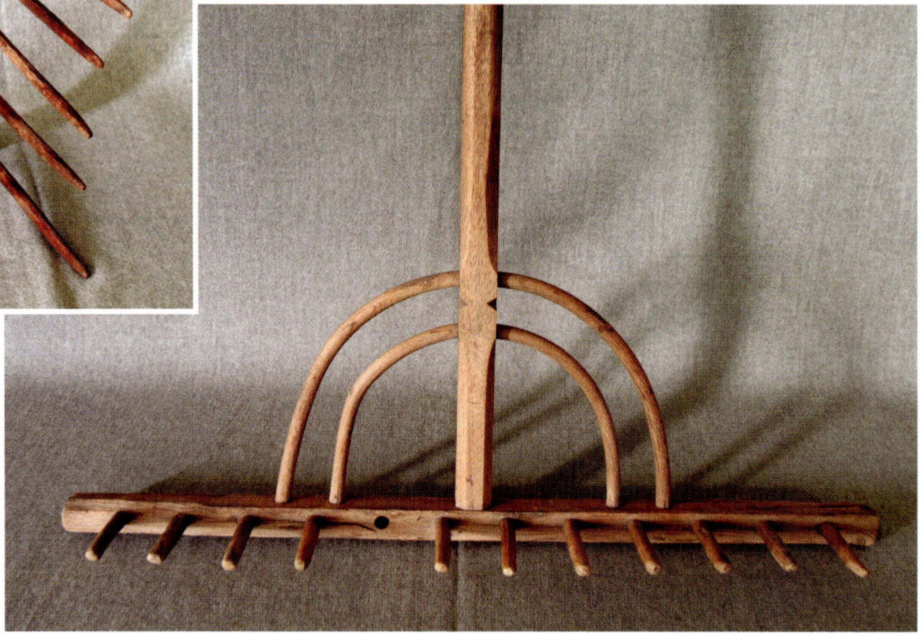

Hay Rakes. The utensils come in all sizes. Mike Emery provides scale for large ones.

Harvesting and Loading Hay. Interpreters at Landis Valley demonstrate the use of rakes and hay forks.

Tumble Rake (*Bartzelbaum Reff* in Pennsylvania Dutch). Horse-drawn, this implement was photographed by Henry K. Landis.

Loading Hay. A Landis Valley interpreter uses a historic hay fork.

Hay Forks. Traditionally made of wood, some are even made from a young tree and its roots. Occasionally they are signed as is the "M. B. Young" example.

Hay Forks.

Work or Hay Wagon with Hay Ladder. The ladder enabled farmers to manually load wagons.
Historic photograph by Henry K. Landis.

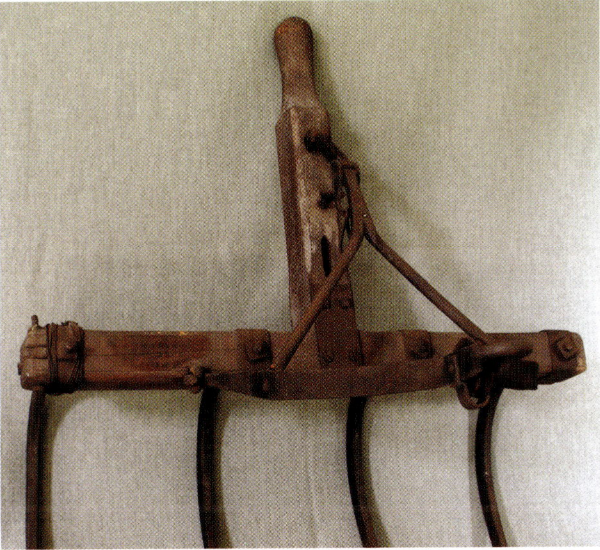

Hay Lifters or Hay Hooks (and details). *Hoi hoake* **in Pennsylvania German, these devices used horse power to load wagons and build hay stacks.** *Historic photographs are by Henry K. Landis.*

Hay Lifters.

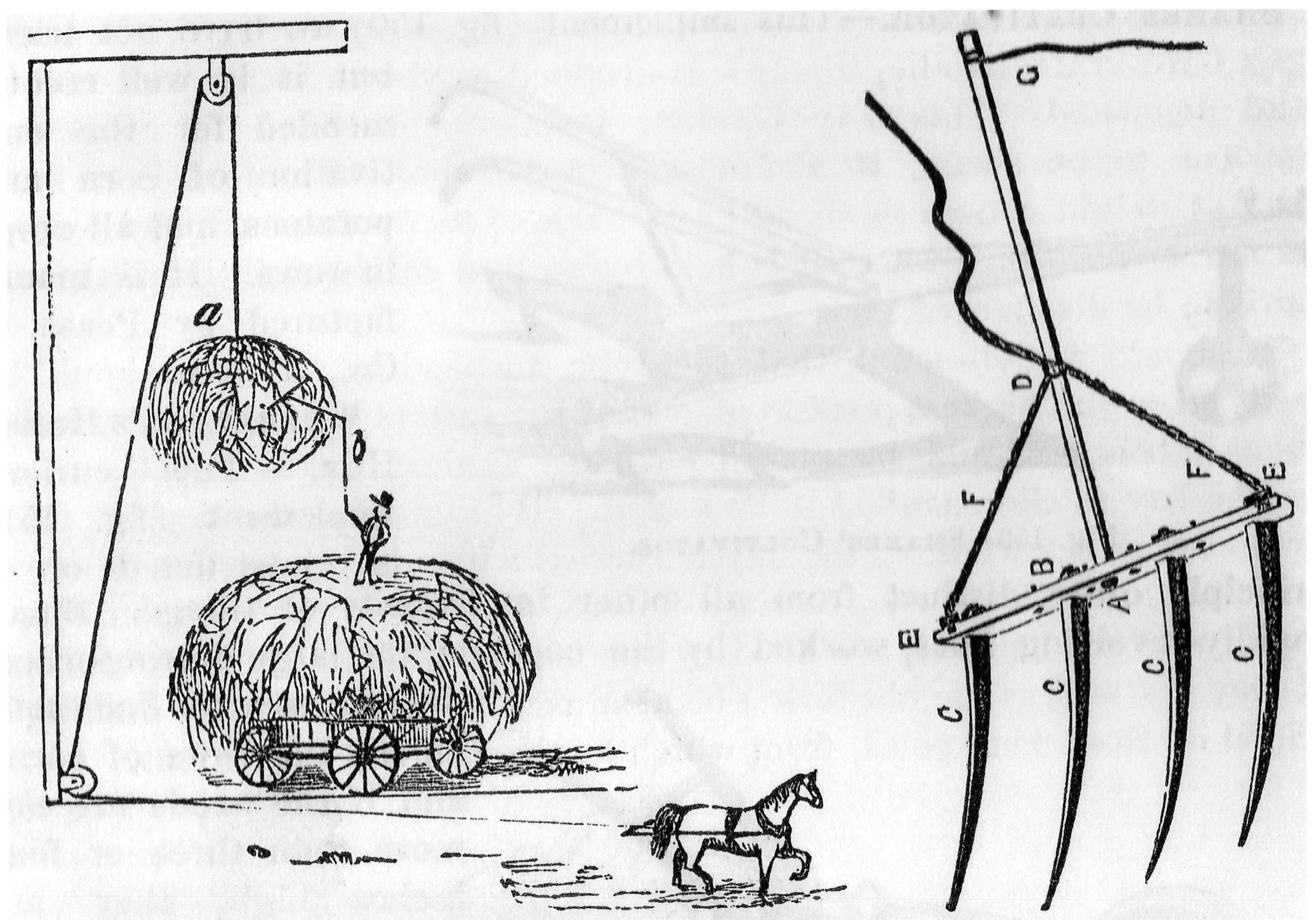

Hay Lifter.

Hay Lifters.

Hay Knives. Used to cut hay from hay stacks after the hay has compacted.

Hay Spades. Like hay knives, these are used to loosen or cut hay from settled stacks.

Bale Hooks. Used in moving hay bales.

Chapter 3

Grain and Potatoes

As the 1985 movie "Witness" opens we are shown a ripening grain field, the wind causing a rippling effect, a device used to establish the rural syntax of the film. Waving fields of grain are symbols of prosperity, as are numerous tall silos. While today we equate grain fields with the mid-west, we must always remember that in the 18th and early 19th centuries, Pennsylvania was commonly considered to be "the bread basket of America" and grains were familiar crops throughout the East. Grain was a staple in the economy of the original 13 colonies and in some form or other retains an important presence in many of our states.

Cereal crops used for bread, the "staff of life," was the primary reason for grain culture. But as man's dietary aspirations developed, so did the raising of grain for animal foods. Today, in our Western culture, far more grain is fed to animals than is directly fed to humans. Five of the six of our major grain crops are old world introductions into the Americas: wheat, barley, rye, oats, and rice, with corn being a crop that is distinctively American in origin. All but rice, which is a warm climate crop, were widely grown throughout 19th century America. Since the 18th century rice has been cultivated in some southern states.

In addition to being a foodstuff, grain has also been very important as a source of alcohol. The new Federal government's first attempt to tax alcohol led to the Whiskey Rebellion (1791-1794). Liquor and beer are primarily grain-based and American legends are wrapped around "corn likker," rye whiskey and Kentucky bourbon. New to the mix is the concept of 'biofuels' – ethanol production from corn intended to help diminish our dependence on petroleum.

Closely allied to grain production as an important form of starch suitable for human and animal consumption, as well as for alcohol production, is the potato, an American plant first appreciated in Europe. The potato became the basis of the Irish economy in the 18th century and, when the crop failed, the ensuing famine and the mass exodus from Ireland would change the population mix of the United States. The potato became the basis of vodka production, the signature liquor of Russia and much of eastern Europe, which eventually came to America. Preparing potatoes for animal food was more labor intensive than some other foods because the tubers often had to be cut up and the potatoes had to be cooked. Additionally, cooked potatoes must be used quickly or they spoil. In the diary of Henry H. Landis (1838-1926) one comes across many references to potatoes and their use as cattle food, always cooked

Opposite: Grain Growing has been pivotal to American agriculture. The lovely lady, surrounded by ripened grain, exhorts, "Come out on the Farm for 'Health and Wealth'" on a postcard sent from Atkins, Iowa, to Cockeysville, Maryland, in 1914.

Near Right: A 1936 prize-winning corn farmer assumes a congruent pose while leaning against a filled corn crib as he gives a testimonial for "Reichards Animal Base Fertilizers."

Far Right: "Yellow Gold" stands in a Pennsylvania field and a little girl watches ducks in a farmyard on a card embellished with the almost universal agriculture symbol of prosperity … a sickle with a sheath of harvested grain, here in 3D.

4th PRIZE—CORN
107.18 Bushels

MARPLE BROS.
Springhouse, Pa.

"We have been using Reichard's Animal Base Fertilizers for several years, and find it exceeds all others that we ever used. The results are most gratifying and encouraging as shown in our 1936 yield.
Clinton Marple

(These boys took 2nd prize the previous year in potatoes, (500.5 bushels per acre) and give Reichard's Fertilizers credit for both yields.)

first. Specialized tools associated with potato culture are seldom recognized today except, perhaps, in areas where potato culture was, or is, dominant. As with grain, mechanization greatly changed the techniques associated with this vital crop. Potatoes are seldom used for animal food today. But where would American popular culture be without French fries and potato chips?

Grain traditionally was hand-cut, a slow and laborious enterprise employing sickles and cradles – tool types often shared with hay harvesting. Mechanization would make the process easier beginning with the reaper, which, for the first time, introduced mass production techniques into grain harvesting. C. H. Wendel in the *Encyclopedia of American Farm Implements and Antiques* has observed that:

> …the reaper revolutionized the harvest, just as its descendant – the grain binder – would do once again in the 1870s. In another 50 years the combine would once again reshape the harvesting of grain. The reaper of the 1830s revolutionized the harvest, bringing to an end the methods going back for centuries. Yet, in less than 100 years, the reaper and the grain binder both came to an end; today, the harvest has been completely mechanized with the combine.

Whole classes of once-familiar farm equipment, from corn huskers and shellers to winnowing baskets, were rendered obsolete and eventually some favored few would acquire "antique" status. Especially desirable are smaller items, which have a sculptural quality and fit within folk art aesthetic popular in the late 20th and early 21st centuries … hence, the great collector interest in scythes and sickles.

Closely allied to grain harvesting has been the milling industry, which historically was a small scale enterprise serving local communities in the age of handcraft. The artifacts of this earlier age are also interesting today for their sculptural quality – especially early hand trucks and millstones. Grain bags and stencils have strong graphic appeal.

Planting grain fields was traditionally done by hand-sowing. Over the years many devices were developed to ease the process and make it more efficient. The derby-hatted sower holds a canvas bucket with a metal frame in which the seeds were held. A. K. Bowers' horse-drawn seed drill, seen in this contemporary advertisement, is similar to one in the Landis Valley Museum. *Sower. Collection: Mr. and Mrs. Michael B. Emery*.

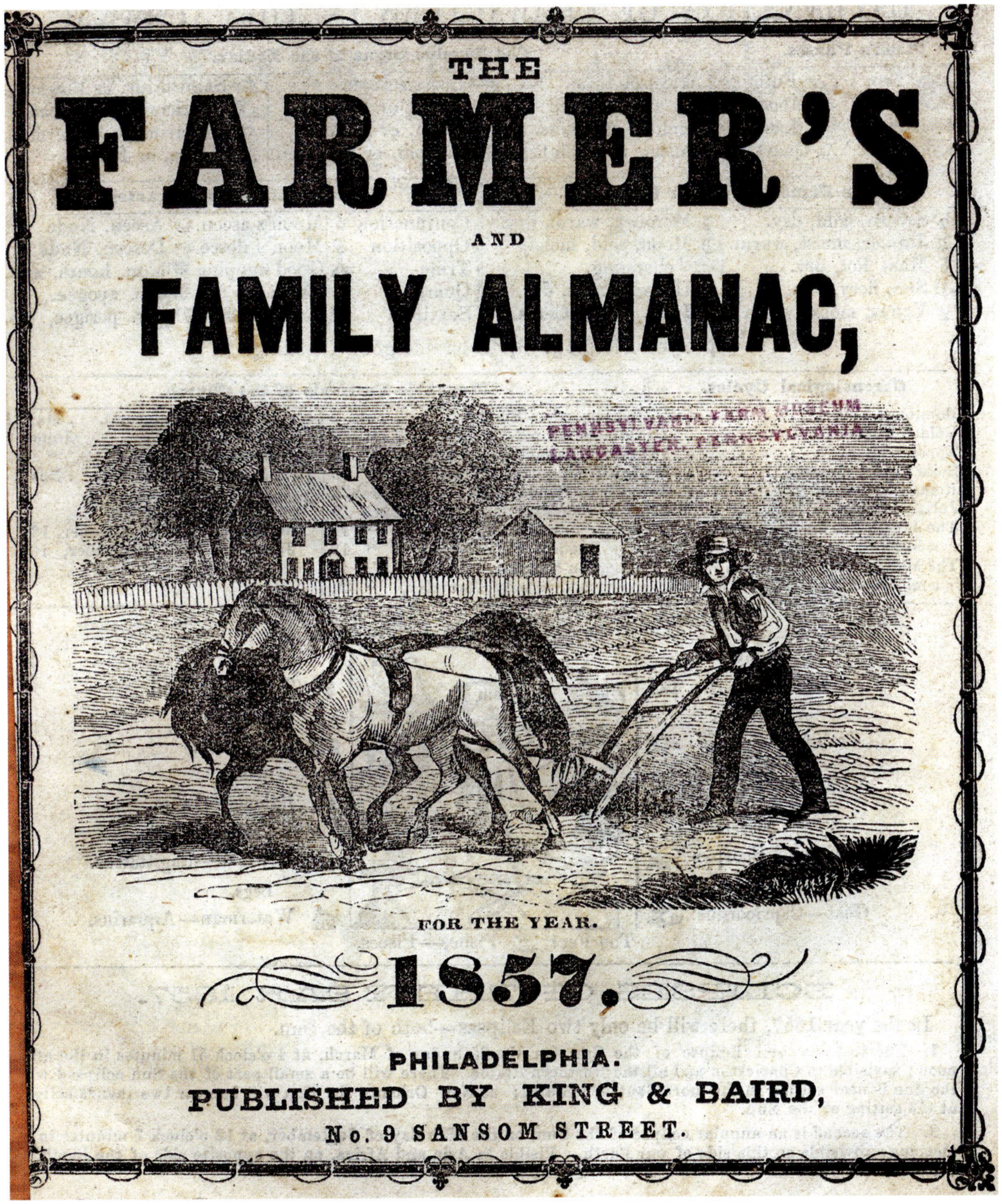

THE

FARMER'S

AND

FAMILY ALMANAC,

FOR THE YEAR.

1857.

PHILADELPHIA.

PUBLISHED BY KING & BAIRD,

No. 9 SANSOM STREET.

A farmer, a team of horses, and a plow are emblematic of American agriculture. There are literally hundreds of variations of plows, both wooden and metal, known to have been used in pre-industrial agriculture.

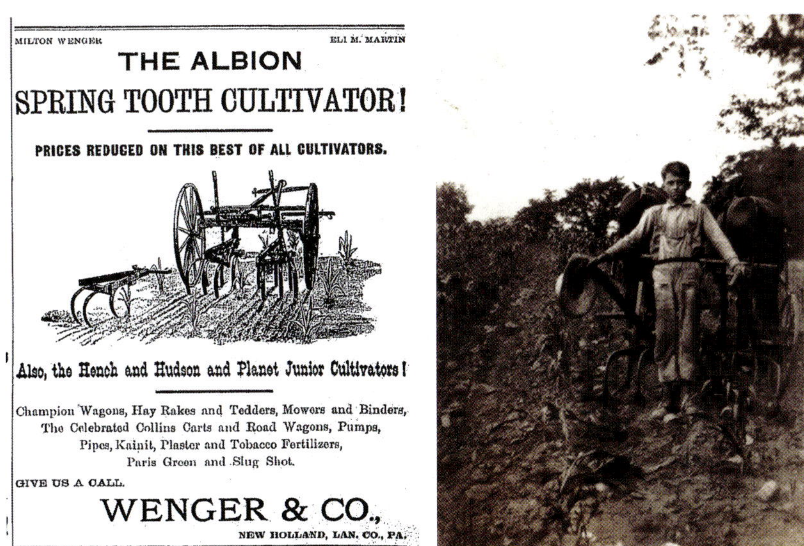

Before the age of herbicides it was necessary to cultivate corn fields, as well as other crops, when the plants were young to keep down weeds. A young farmer poses with his horse-drawn cultivator early in the 20th century. *Collection: Mr. and Mrs. Michael B. Emery*.

A straw-hatted farm worker heads to a field armed with his scythe. He also carries a water or cider jug to quench his thirst during his hot, physically demanding, work. *Collection: Mr. and Mrs. Michael B. Emery*.

Cutting grain, as with hay harvesting, was commonly done with sickles or other hand tools prior to the horse drawn machine. In the 19th century woodcut we see stylized farm workers harvesting a grain field. The photograph, c. 1930, is no doubt staged to show how the crop was cut historically. Note the wooden rake in the foreground.

Grain reapers, pioneered by Cyrus Hall McCormick in 1831, revolutionized grain harvesting by using animal power to substitute for human power. The machine cut grain stalks and neatly stacked the harvest, which then could be more easily tied into sheaves.

The western scene shows two reapers at work. The horses to the left are probably pulling a load of sheaves. The rye harvest field is in Wayne County, Pennsylvania. Both images date from early in the 20th century. *Field of rye, Collection: Mr. and Mrs. Michael B. Emery.*

Proud of their successful harvest, a Lehigh County, Pennsylvania, farm family poses amidst binder-stacked wheat. On many farms the whole family helped bring in the harvest. A reaper drawn by three horses stands to the right. *Collection: Mr. and Mrs. Michael B. Emery*.

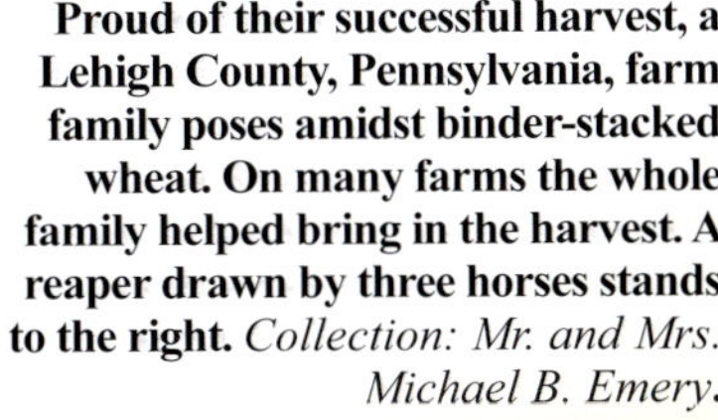

Harvesting grain, east and west, with animal power. Where mountains and hills are visible and fields are smaller, rigs were generally drawn by two to four horses. On California's vast fields, huge teams of mules propelled harvesters. Eastern scene, *Collection: Mr. and Mrs. Michael B. Emery***.**

The transition from horse power to the gasoline-powered tractor was gradual among conservative farmers. Machinery, reaper and binder designed to be pulled by animals, were often adapted to the machine age. *Collection: Mr. and Mrs. Michael B. Emery***.**

Once grain was harvested the heads had to be separated from the straw. The ancient way of doing this was to beat the heads of grain loose and separate the chaff using a flail, as seen in this mid-19th century view. Threshing machines made the process easier. Here two Amish men feed hand-tied grain sheaves into a thrasher powered by a stationary engine. Today modern combines do all of the grain harvest tasks in one – cutting, threshing, and winnowing – followed by trucks to take the grain and its by products off. *Collection: Mr. and Mrs. Michael B. Emery*.

Once grain was flailed or threshed it had to be winnowed to separate the cereal from the chaff. Anciently this was done with a winnowing basket and the wind; by the late 18th century most farms used a fanning mill. The grain and attached bits of straw was shoveled in and clean grain emerged. Most mills were operated by boy power, although large ones would employ horses or mules. Photographed in the early 20th century in Berks County, Pennsylvania, the process was recorded by H. Winslow Fegley: *Courtesy: Schwenkfelder Library and Heritage Center.*

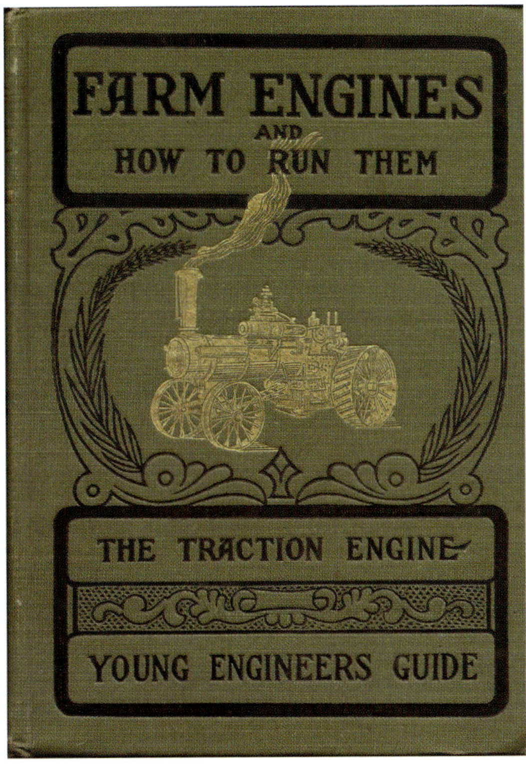

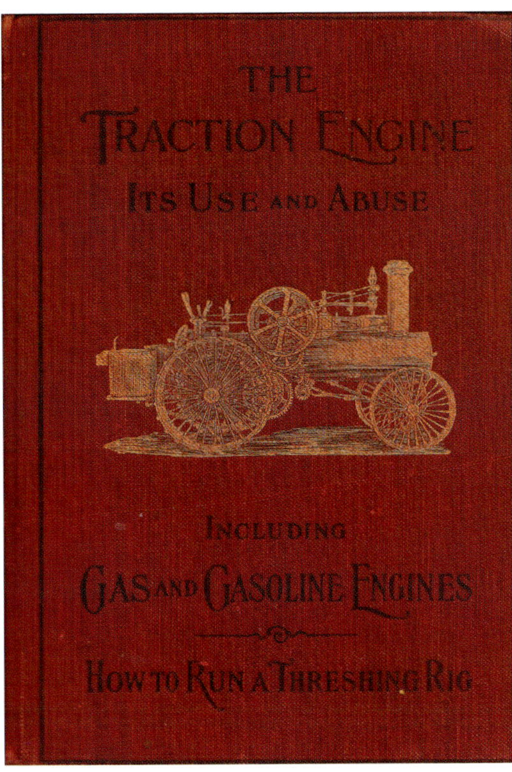

THE ONLY GENUINE
Geiser's Patent Self-Regulating

—GRAIN—
Separator, Cleaner and Bagger
AND THE
PEERLESS,
TRACTION,
—AND—
DOMESTIC STEAM ENGINE
MANUFACTURED BY THE
Geiser Manufacturing Co.
For particulars call on or address
J. M. STAUFFER, Agt.,
my15 11t New Holland, Lan. Co., Pa.

When steam and gasoline power became popular many books were published to instruct farmers and "Young Engineers" on how to use and service them. An early power application relieved generations of boys and animals from operating fanning mills. *Collection: Mr. and Mrs. Michael B. Emery*.

Most early steam engines were stationary. This grand piece of machinery would be taken into the field and could be attached to a thresher, or it might be used to operate an elevator to more easily store hay.

Milling grain was important to the local economy. Mills were everywhere in grain and fiber-growing areas. Landis Valley Museum co-founder, Henry K. Landis, holds an unidentified object as he poses amidst a collection of his prized millstones, ***circa*** **1940.**

A monster-sized steam tractor draws huge equipment across a California grain field. Back east, a small gasoline-powered engine helps with the harvest on a more modest scale. *Collection: Mr. and Mrs. Michael B. Emery*.

When ready to be harvested potatoes would be exposed with a potato-digger – a specialized plow-form. This New York model dates from circa 1875. No example of this tool is now in the Landis Valley Museum. *Collection: Mr. and Mrs. Michael B. Emery*.

After the plow uncovered the potatoes the tubers had to be harvested by hand. On many farms entire families would be involved. In this scene at Landis Valley photographed by Henry K. Landis, men and boys, probably hired hands, do the work. A wagon to take the harvest from the field is in the background.

Once harvested, potatoes needed to be sorted and bagged. By the late 19th century horse-drawn equipment helped with the task. *Collection: Mr. and Mrs. Michael B. Emery*.

From the Landis Valley Collection

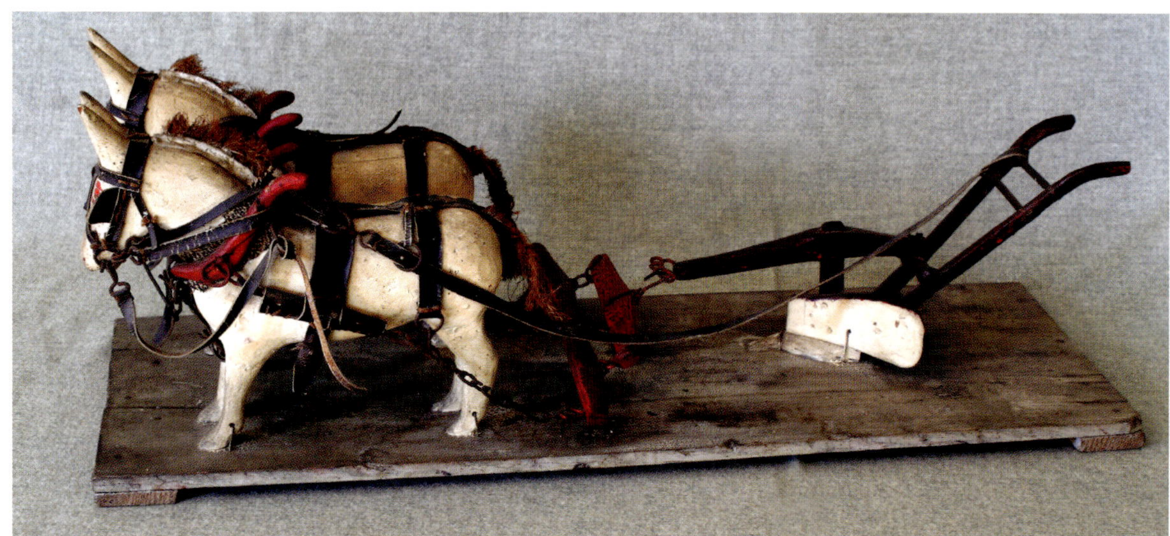

Model Plow and Team

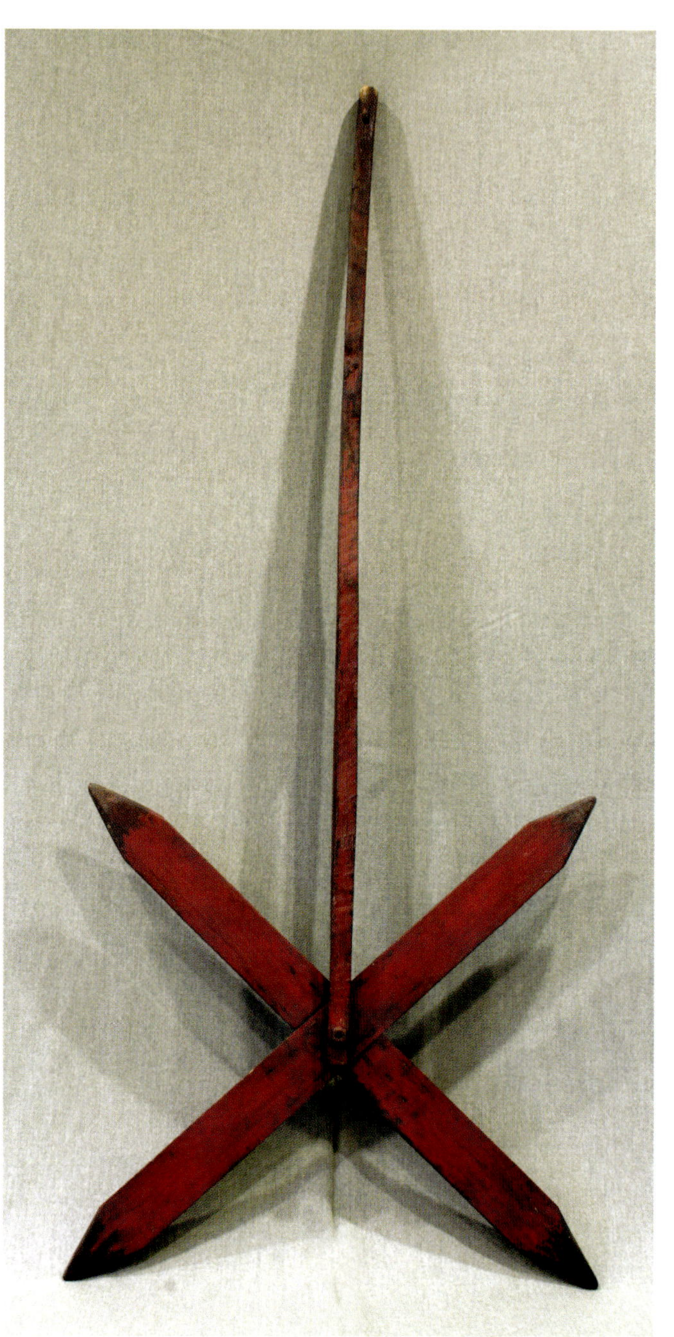

Row Marker. Used after the soil was prepared.

Plows and Details. The most important of farming tools can be traced to antiquity and they were made of wood and metal. Early ones were pushed and/or pulled by humans. These examples were pulled by horses or oxen. The "Oliver Chilled Plow" was high-tech in the late 19th century. With the coming of tractors gang plows that could open multiple furrows became ever more common.

Plows.

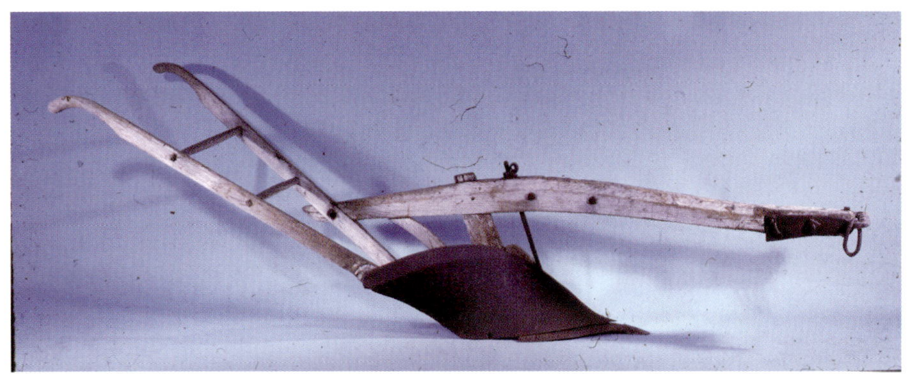

Plows.

Plowing at Landis Valley.

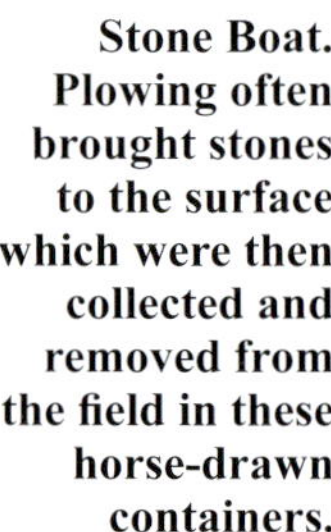

Stone Boat. Plowing often brought stones to the surface which were then collected and removed from the field in these horse-drawn containers.

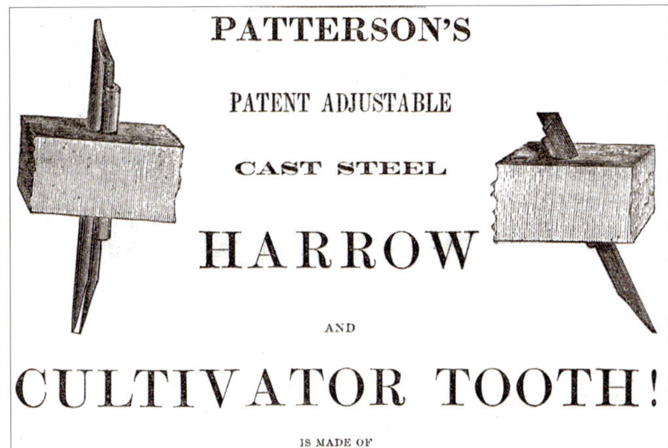

Harrows. Used since ancient days for leveling and smoothing soil. The first harrows are alleged to have been small trees dragged across fields.

Disk Harrowing at Landis Valley.

Germination Tray. Tested the viability of seeds.

Grain Seeder. Turning the crank sent a cascade of seeds in all directions

Planters. Used to precisely space seeds. Although used most commonly for the planting of corn, most could be used for multiple varieties of seeds.

Corn Planter.

Seed Drill. A horse-drawn implement to precisely plant grain fields.

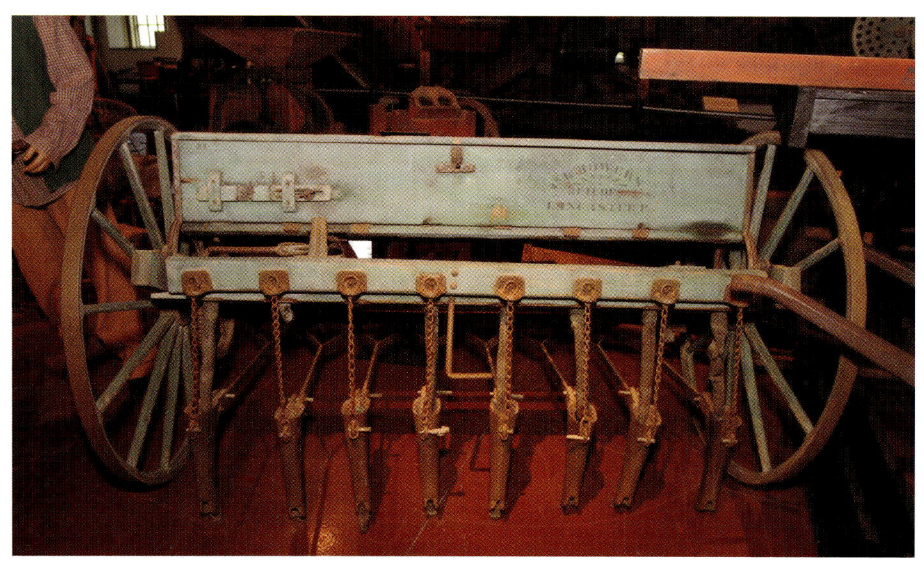

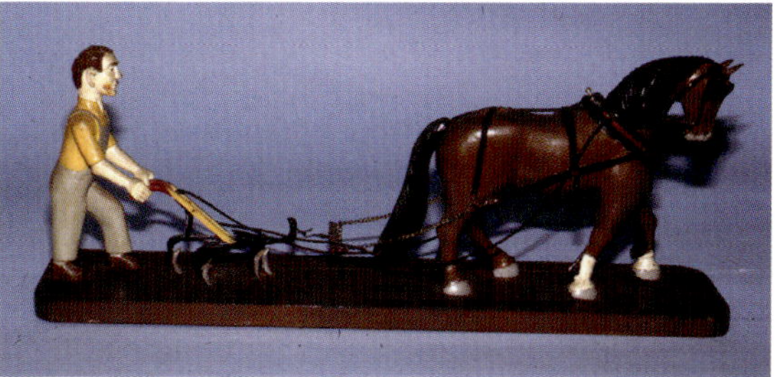

Model of a Cultivator.
The principle horse-drawn weed controller.

Row Closer. Used to cover seeds after plowing, harrowing, and planting.

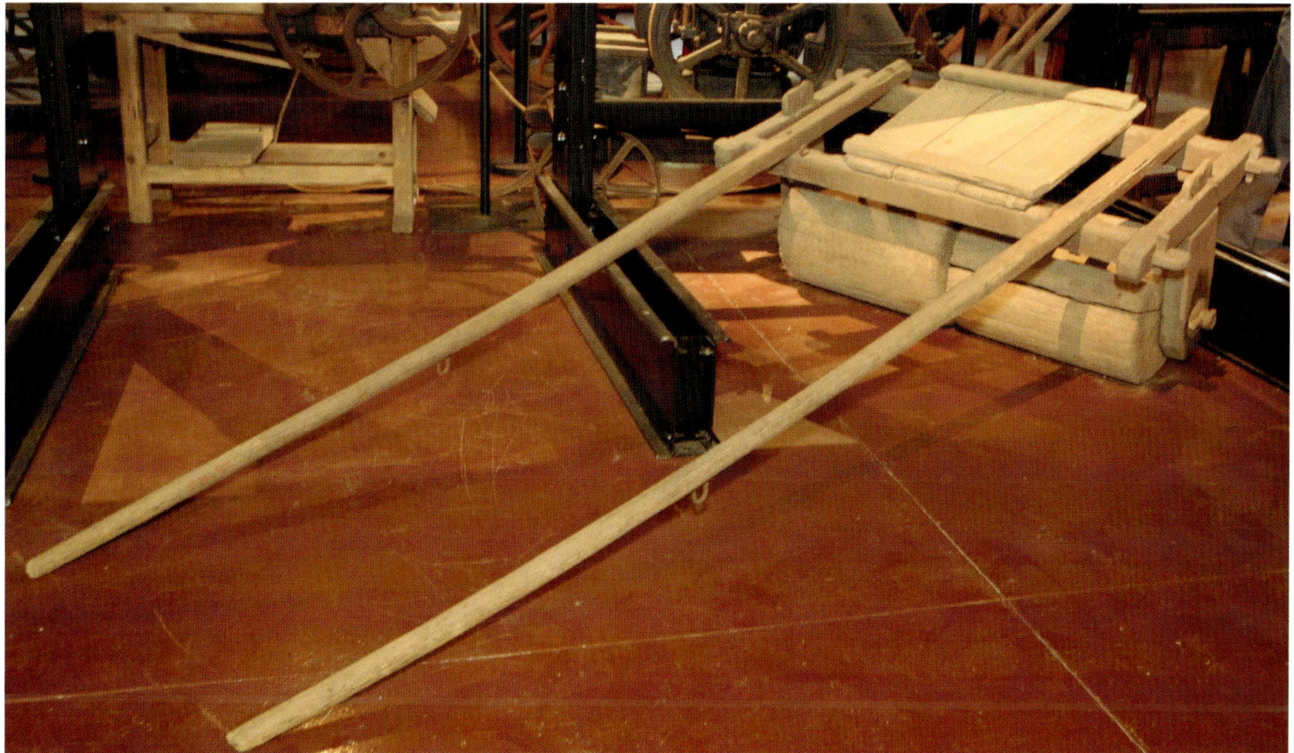

Land Roller or Cultipacker. Employed to level seed beds. In winter they were used to pack snow down on roads to ease the use of sleds.

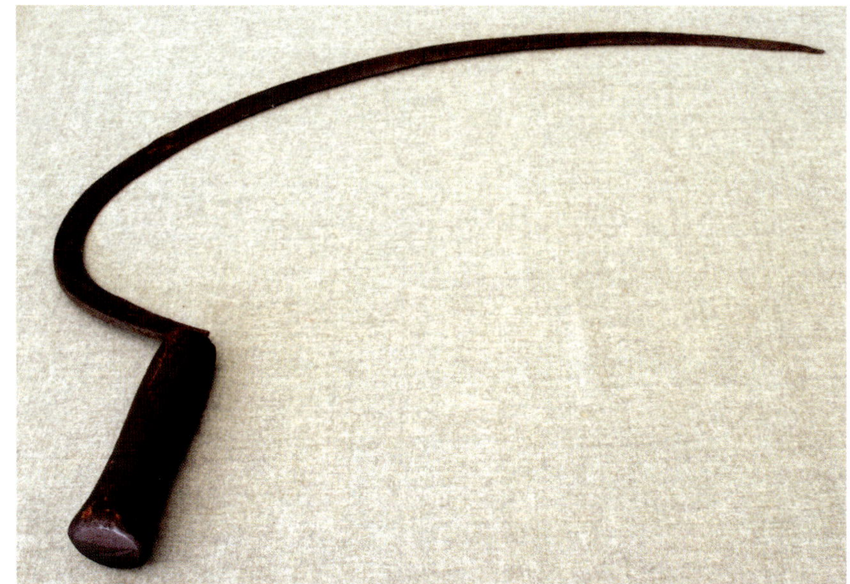

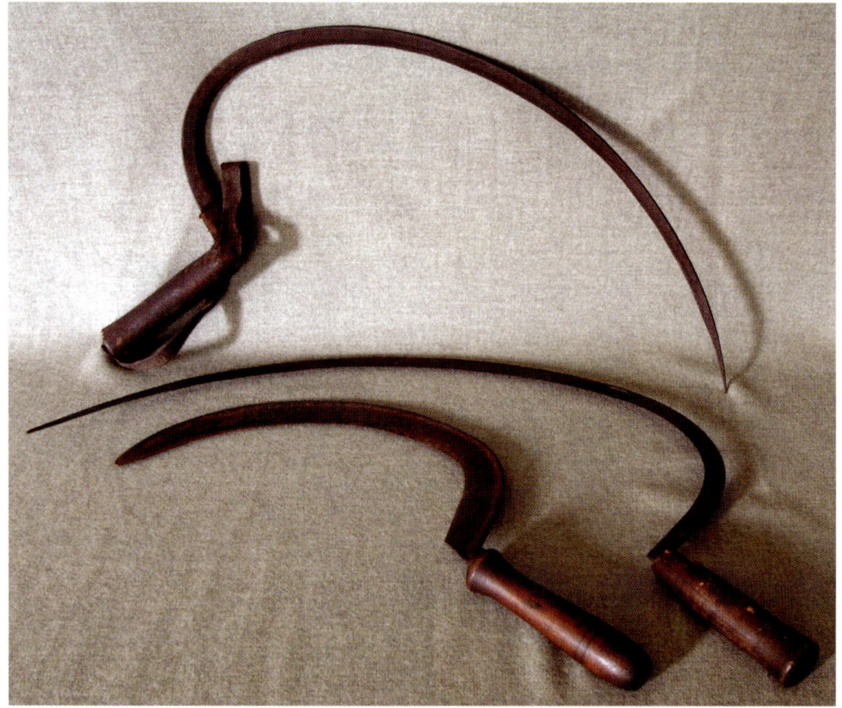

Sickles and Crooks. Sickles as grain harvesters are ancient in origin. The crooks use the natural form of saplings.

Grain Cradles. Used to cut and organize grass-like grains.

Corn Knives. Used to cut corn stalks in the field.

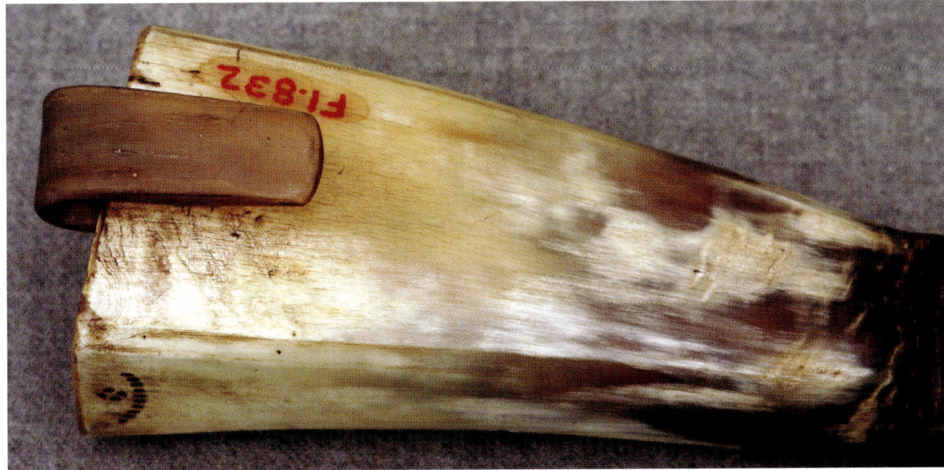

Whetstone Holders. Usually made from animal horn and hung on the farmer's belt, these containers held whetstones used to sharpen tools in the field.

Dengelstocken **and Hammers. Portable anvils, the** *dengelstock* **was driven into a convenient tree and, with the hammer, allowed a farmer to fix his cutting instruments if they hit a rock while in the field.**

Work Bench with a Stationary ***Dengelstock*****.**

Corn Cutter. A mechanical harvester.

Reaper (Model). The harvesting machine that revolutionized agriculture.

Flail. The ancient hand tool used to separate grain from straw. In some places it was used until *circa* 1900.

Threshing Scoop. Helped separate threshed grain from bits of straw.

Thresher (model). Mechanically separated grain from straw. In common use until the 1950s when it was largely replaced by the combine.

Threshing Fork. Used to feed harvested grain into a thresher.

Threshers

Bale Press. Created bales from threshed straw and mown hay.

Fodder Choppers or Shredders. Reduced corn husks and stalks into feed-sized pieces.

Husking Pegs. Used in corn husking.

Corn Shellers. Removed kernels from the cobs. They were developed in the 1850s and took many shapes and forms.

Corn Shellers.

Grain Mills. Used on farms primarily to grind animal feed; however, artisan millers continue to use them to make cornmeal today.

Cornmeal Grinding at Landis Valley.

Wooden Shovels or Scoops. A local craft. Here we see shovels in process as well as the finished products. Wooden scoops were used in handling grain in a dusty environment where a spark from a metal tool could spark an explosion.

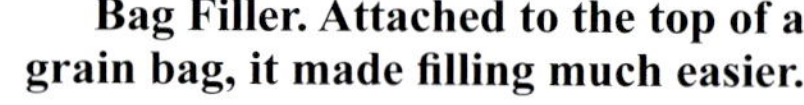

Bag Filler. Attached to the top of a grain bag, it made filling much easier.

Fanning Mills or Grain Cleaners. These allowed the farmer to clean weed seeds and other particular matter from his grain.

Hand Truck. Very important in moving grain sacks, many early ones were locally made, often of wood.

Grain Sacks. Made of rough linen fiber, they were highly prized and refilled often.

Grain Sack Stamp. Used to mark the owner's name on a grain bag. Carved backwards, when printed you could easily read that this 1837 stamp was used by Samuel Wise.

Markings on Grain Bags. The H. H. Landis bag belonged to Henry Harrison Landis, the father of the founders of the Landis Valley Museum.

Potato Harvest Basket. Made of split oak, this single-handle basket was pulled along rows during harvest.

Potato Sorter. This locally made implement graded potatoes according to size.

Peerless Steam Tractor. Heavy, massive, steam tractors were the most majestic pieces of farm equipment in use from the late 19th century to the early 20th century.

You May Think That the Advertisement of the

New Holland Gasoline Engine

Was not so much of an advertisement of the Engine as of the Pump Fixtures and Connections. Well, we have decided to show you our Engine mounted on our 4-wheel truck.

The above shows our 1½ H. P. Engine mounted on a 4-wheel truck. This truck is strongly built, painted, striped and varnished. The sills are made of wood and the wheels of cast iron. This makes it very convenient to move the Engine from place to place. You can move it easily to the house to run the wash machine, butter churn, water pump, cream separator and then again move it to the barn to run your corn sheller, horse clipper, etc., etc.

THE NEW HOLLAND WOOD SAW.

Here we have a simple and cheap Wood Saw which can be run by the above Engine for cross-cutting and ripping. For ripping boards we furnish a flat table provided with an adjustable gauge. For full particulars and prices, address or call on—

New Holland Machine Company,

NEW HOLLAND, Pennsylvania.

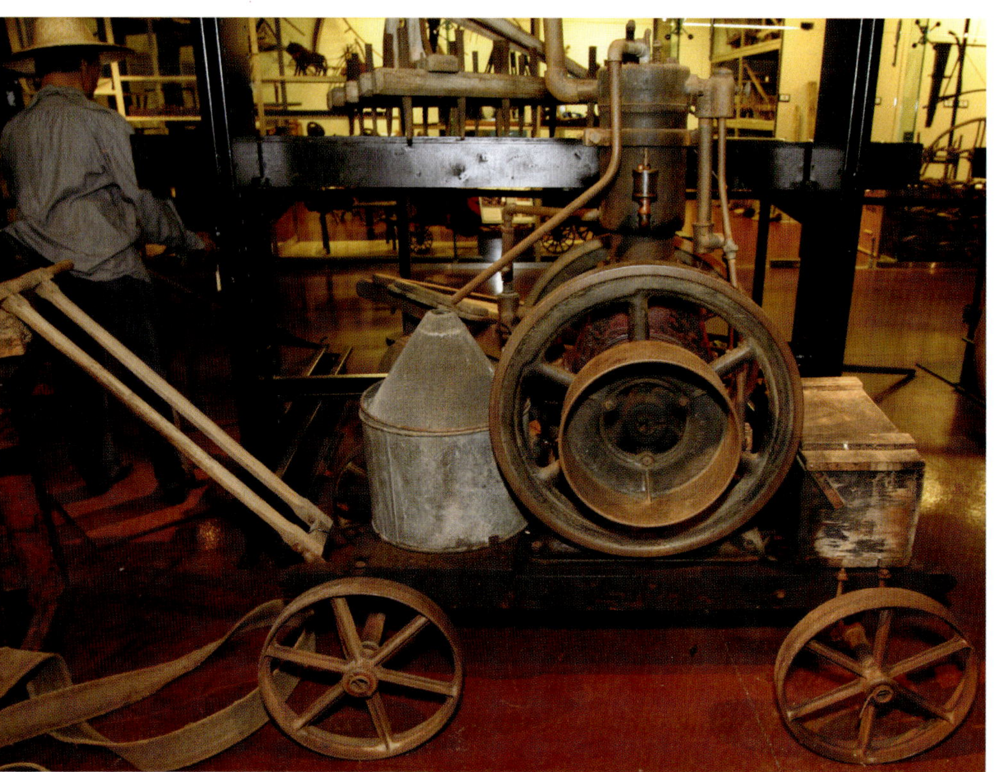

Stationary Gasoline Engines. These were harbingers of the agricultural revolution that the internal combustion engine fomented. Stationary engines could power dozens of other pieces of equipment, which had relied on human or animal power. New Holland Farm Machinery Company was a leading manufacturing of these devices as well as an extensive line of other agricultural machines.

Frick Gasoline Tractor, 1917. This machine is one of the earliest of the lighter weight gasoline tractors that were so important to the evolution of farming. It is most impressive to see such machines in historic out-of-doors photographs taken at Landis Valley.

Early Case Tractor.

CASE
INTERNATIONAL
CASE
WOLF

Chapter 4

Tobacco

Ever since tobacco was introduced into European society in the 16th century, its usage has been controversial. While most had enjoyed the euphoria it could induce, a minority denounced the "filthy weed." In 17th century Maryland, growers called it "Sot Weed." Nevertheless, growing, trading, and exporting tobacco became central to the economy of large parts of the American South. John Barth's classic novel about the Chesapeake Bay region in the 17th century was entitled *The Sot-Weed Factor*.

Before proper manuring practices were understood and employed tobacco was denounced as an "exhauster of soil." But land was cheap and the quest for new tobacco soils opened the inland upper south to agriculture.

By the 19th century tobacco became a major crop ranging from the Connecticut River Valley in the North down through North Carolina. Originally tobacco was smoked in pipes or ground into a powdery snuff. Early pipes had very small bowls because tobacco was expensive. As tobacco became cheaper, pipe bowls enlarged. Snuff was sniffed or snorted into the nose or placed between the gums and the upper lip. Beginning in the 17th century elegant snuff boxes – often of precious materials and gem encrusted – became popular. Ordinary folk made do with boxes made of wood, horn or base metals.

The development of chewing tobacco and cigars further spread the use of tobacco, but bringing tobacco use to its zenith was the rise in the use of cigarettes first popularized around World War I. While some women discretely used snuff and a few radicals would smoke cigars, tobacco usage among women only really became widespread with the cigarette. By the 1920s "smart set" women smoked; by the 1950s many women did. The fortunes of different regions in the tobacco area changed as tastes changed. Tobacco grown in the north was best suited to cigars; that grown in Virginia, Maryland, and North Carolina worked best in cigarettes.

Tobacco, known botanically as *Nicotiana,* **is named for French diplomat Jean Nicot (1530-1600). The original tobacco brought into Europe and first cultivated in America was** *Nicotiana Tabacum*. **"A young Virginia maiden," as described on the reverse of this 1950s postcard, poses next to a fully mature tobacco plant. In most areas the tobacco would have been headed, or "topped," to prevent flowering – a hand operation using a sharp knife. Some nicotianas are raised just for blossoms, especially** *Nicotiana alata*, **commonly called flowering tobacco, known for its sweet nocturnal aroma.** *The Cottage Collection.*

In Central Pennsylvania growing tobacco became very important economically. The cities of York, Lancaster and Reading sported tobacco warehouse districts and a lively cigar manufacturing industry devoted to inexpensive and middle-market products. Pennsylvania tobacco was especially suited to cigars, but cigarette grade tobacco could also be grown. Tobacco was a favored crop because it could be so profitable and it was often proclaimed "the mortgage burner" crop. Henry H. Landis (1838-1926) kept a detailed diary in which he always capitalized "Tobacco." He never extended the same honor to corn, wheat, or hay.

Raising tobacco was, and remains, extremely labor intensive. In today's regional agricultural picture it is mostly raised by the Amish whose large families can provide the required hands. While some Amish men smoke cigars or "stogies," cigarettes are generally forbidden as being too modern.

In late March or early April dust-fine tobacco seed had to be planted in protected, very well tended beds. Young tobacco plants are vulnerable to all kinds of weed infestations as well as insect and disease problems. To counter these threats a new process emerged early in the 20th century whereby the tobacco beds were steamed or "cooked" to sterilize the soil.

A field of tobacco or a "tobacco patch" grows in Lancaster County, Pennsylvania. The long narrow building in the background (center) is a tobacco barn. This postcard was mailed in August 1941 by a couple "… taking a trip through Lancaster County" to a friend in Allentown. *The Cottage Collection*.

What's the hurry? Does the racing Amish man not want his picture taken or did he need to answer a call of nature? Two mules draw a tobacco planter. Two boys plant seedlings previously raised in nursery beds. A tank provides water to each new plant.

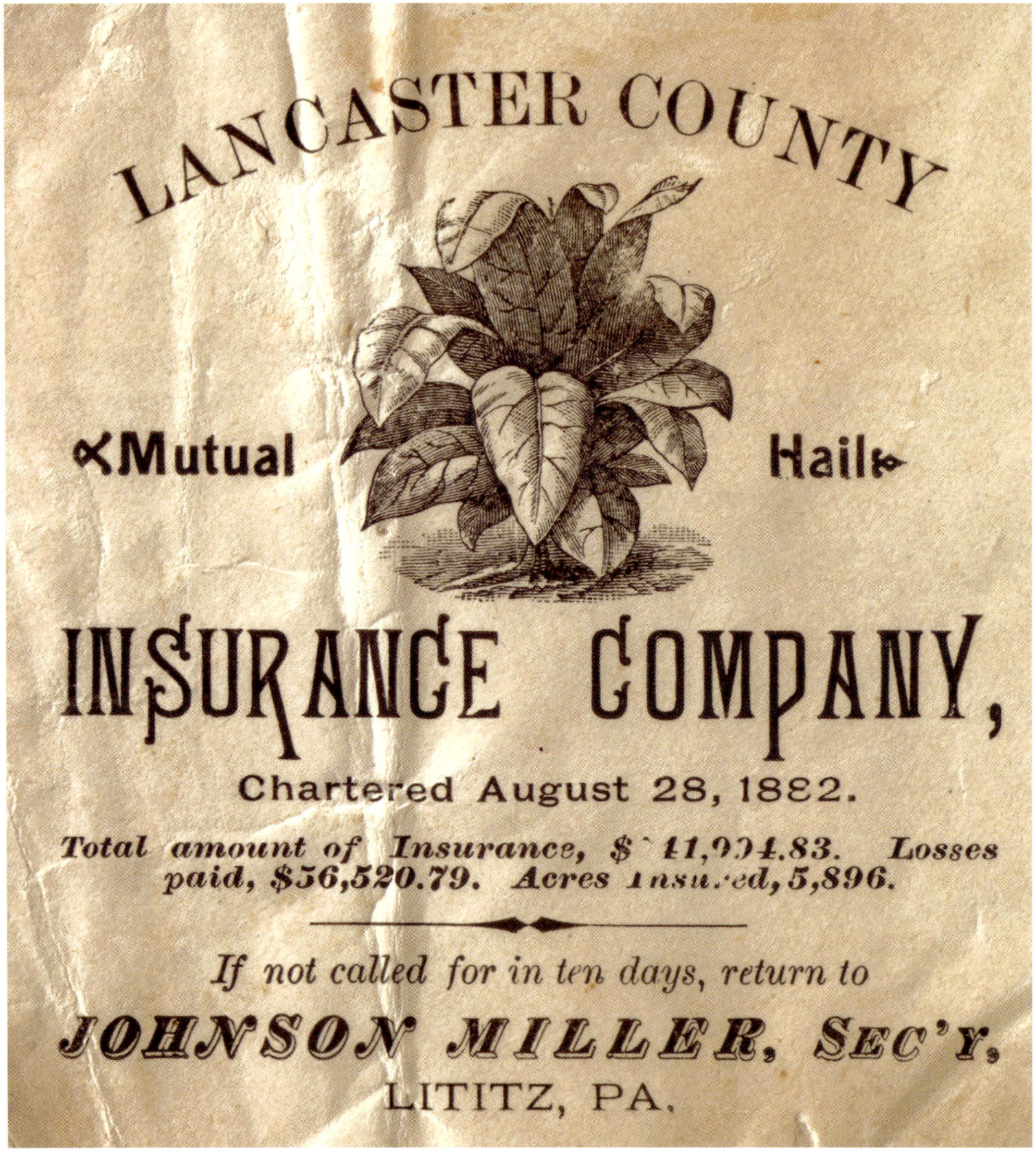

LANCASTER COUNTY

Mutual Hail

INSURANCE COMPANY,

Chartered August 28, 1882.

Total amount of Insurance, $[illegible]41,9[illegible][illegible].83. Losses paid, $56,520.79. Acres Insured, 5,896.

If not called for in ten days, return to

JOHNSON MILLER, SEC'Y,

LITITZ, PA.

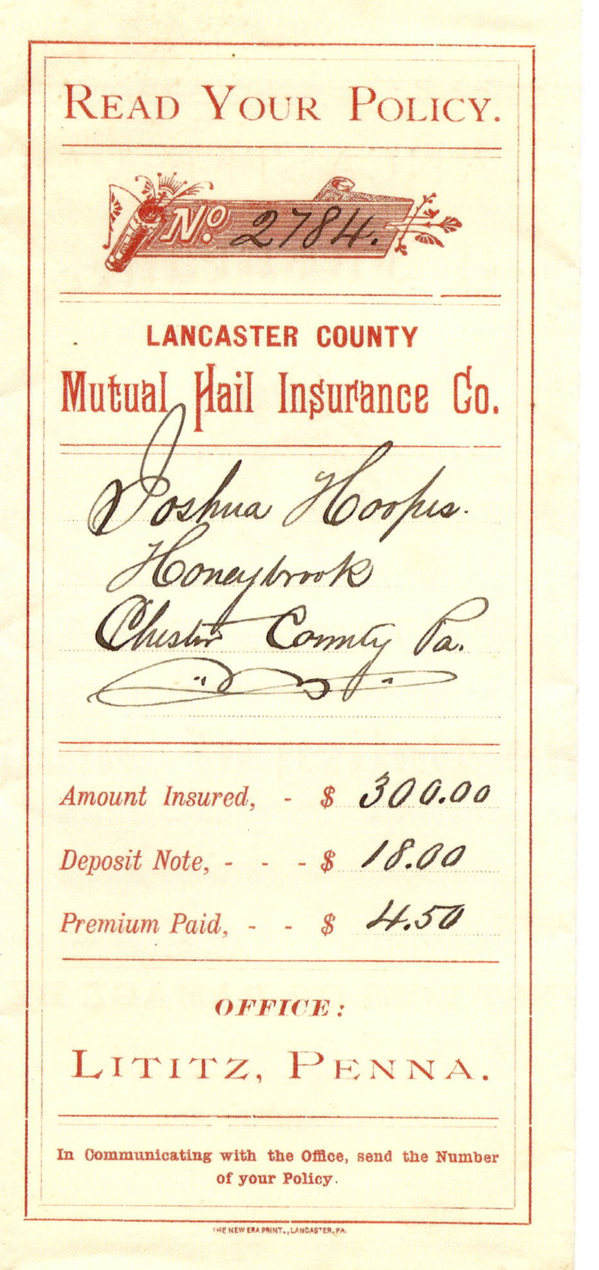

READ YOUR POLICY.

No. 2784.

LANCASTER COUNTY

Mutual Hail Insurance Co.

Joshua Hoopes.
Honeybrook
Chester County Pa.

Amount Insured,	$	300.00
Deposit Note,	$	18.00
Premium Paid,	$	4.50

OFFICE:

LITITZ, PENNA.

In Communicating with the Office, send the Number of your Policy.

THE NEW ERA PRINT., LANCASTER, PA.

Hailstorms are a major problem for tobacco growers as hailstones can badly damage plants leaves. Local insurance companies helped to protect the farmers. The late 19th century policy holder, Joshua Hoopes, was the great-great-grandfather of author Mike Emery. *Collection: Mr. and Mrs. Michael B. Emery*.

In late May the crowded plants, literally in the thousands, could be transplanted into the fields. Horses or mules were used to pull a specialized rig called "a tobacco planter." One or two farm hands sat on this rig close to the ground, placing the plants that had been harvested from the starting beds. An animal-powered water tank followed the transplanters. Once planted tobacco requires constant care. Each tobacco "patch" (we might call them fields) had to be hand-hoed and –weeded. Farmers had to be especially alert for Tobacco worms, which could be very destructive. Poisons could not be used as a control element and children were often paid to pick worms … paid per larva they captured. Tobacco also had to be "suckered" (its side shoots removed) and "topped," to prevent the plant flowering. Hand work all. Topping was done in late July and August.

Harvesting tobacco on the farm of Henry H. and Emma C. Landis was photographed by their son, Henry K., circa 1900. Newly cut tobacco awaits spearing. Four workers pose with plants arranged on laths. An almost fully laden tobacco wagon awaits a final load.

Later in August or early September tobacco was harvested with each stalk cut by hand using a specialized long-handled shear. Next, the stalks were threaded onto lathes, which would be hung on special wagons, keeping the hanging plants suspended. It was important that the leaves not be damaged because that would hurt the value of the tobacco. Once removed from the fields, the tobacco would be transferred to a special barn with slatted or louvered sides where it would air-dry until January when the next process was begun. In "stripping," the dried or "cured" leaves were stripped from their stalks, sorted by size and quality, and finally baled. The work was often done in the basement beneath the tobacco barn, which was customarily heated by a stove.

For the farmer, the best time of the year – usually March or April – was when the tobacco was sold. Weather and ensuing quality and demand determined the price. Sometimes there was great joy. At other times other emotions surfaced. But now it was planting time for the new crop.

The Plain people are among the last tobacco farmers in central Pennsylvania. In this 1970s view Amish and Mennonites harvest the crop. The woman in the foreground spears tobacco plants onto a lath. *The Cottage Collection*.

Chester County, Pennsylvania, farmer Joshua Hoopes (1847-1927), left, and his son, Park (1884-1927), on the wagon, along with a relative, Bill Bailey, pose with harvested tobacco about 1900. The detail of the tobacco wagon is especially fine. *Collection: Mr. and Mrs. Michael B. Emery*

A mule-drawn tobacco wagon loaded with fresh cut plants stands in front of a tobacco barn. Open slats in the barn allow air to circulate freely to encourage even drying of the leaves. *The Cottage Collection*.

The tobacco being hung in the tobacco shed or barn at Landis Valley. Late summer and early fall were especially busy on the farm. A steam engine powers a threshing machine. In the background the long tobacco shed is filled with drying leaf. Photographs are by Henry K. Landis.

A Conestoga Wagon laden with bales of tobacco arrives at a tobacco warehouse in Lancaster City around 1870, at a time when this was clearly a rare sight. Note the bells above each team. Everyone wanted to be in the picture recording the historic event.

Tobacco was ultimately made into snuff, cigars (or segars) or chewing tobacco. The Demuth firm still exists as a retailer of tobacco products in Lancaster. One of its last warehouses is owned by the Demuth Foundation, which is devoted to the memory of a family member, artist Charles Demuth (1883-1935)

J. H. OSTERMAYER,
MANUFACTURER OF SEGARS
AND DEALER IN
TOBACCO, SEGARS, SNUFF,
PIPES, &c.,
No. 68 NORTH QUEEN STREET,
LANCASTER, PENNA.

134 DIRECTORY OF LANCASTER COUNTY.

1770 ESTABLISHED 1770

H. C. DEMUTH,
Successor to JACOB DEMUTH,
MANUFACTURER OF
DEMUTH'S CELEBRATED SNUFFS
AND DEALER IN ALL KINDS OF
CHEWINC TOBACCO,
SUCH AS
NAVIES, CAVENDISH, CONGRESS,
SPUN ROLL, ORINOCO TWIST, ROUGH AND READY, LIGHT PRESSED,
PLAIN CAVENDISH, DIAMOND TWIST, &c., &c,
SMOKING TOBACBO in Endless Variety.
PIPES OF ALL KINDS,
Meerschaum, Vienna, Bead and all other kinds of Bowls, Briar Wood, Rose Wood, Box Wood, &c., &c., at Philadelphia Prices.
H. C. DEMUTH,
No. 49 EAST KING STREET, Lancaster, Pa.

From the Landis Valley Collection

HENRY H. LANDIS,
DEALER IN
Leaf Tobacco,
LANDIS VALLEY P. O.,
LANCASTER CO., PA.
F. J. FAESIG, NATIONAL JOB PRINT, 54 NORTH QUEEN ST.

Business card for Henry H. Landis' tobacco business.

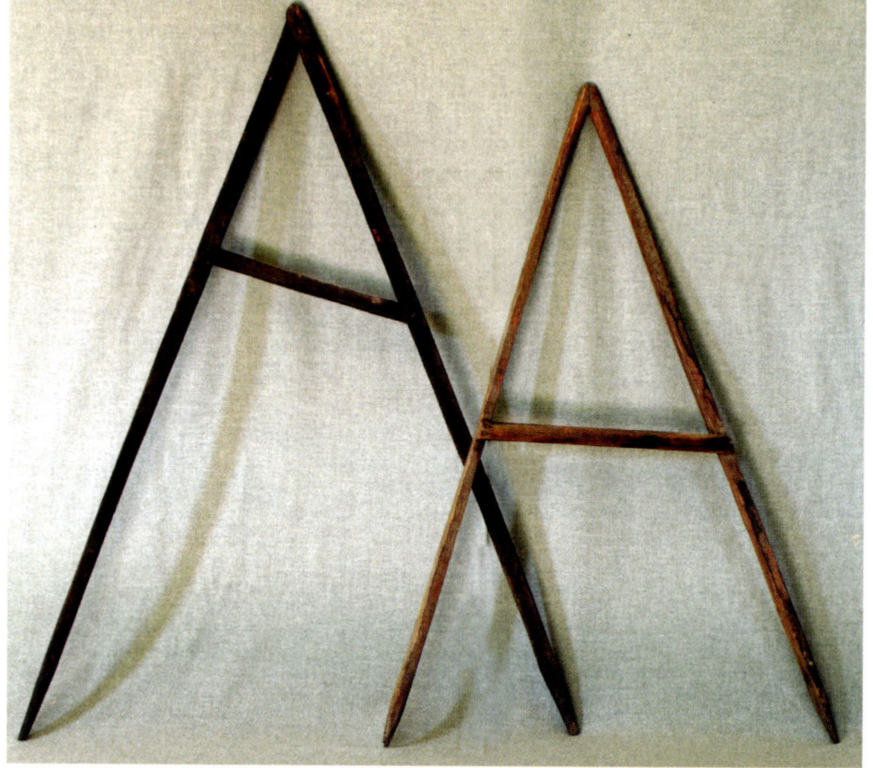

Row Spacer. Used to properly distribute plants.

Tobacco Seeder. Tobacco seeds are extremely fine so a specialized tool is required for effective planting.

Tobacco Transplanter.

Tobacco Hoe. The most common cultivating tool.

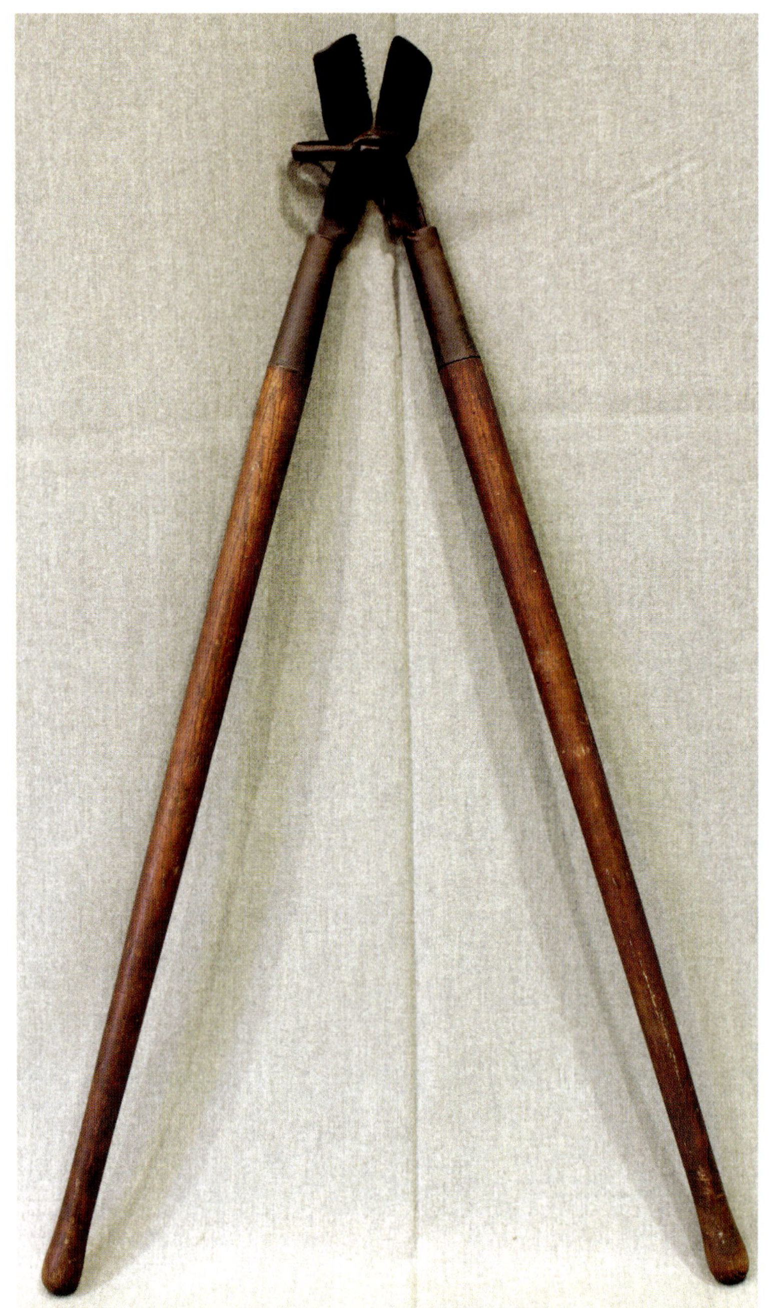

Harvest Shears. Used for cutting the mature tobacco plant.

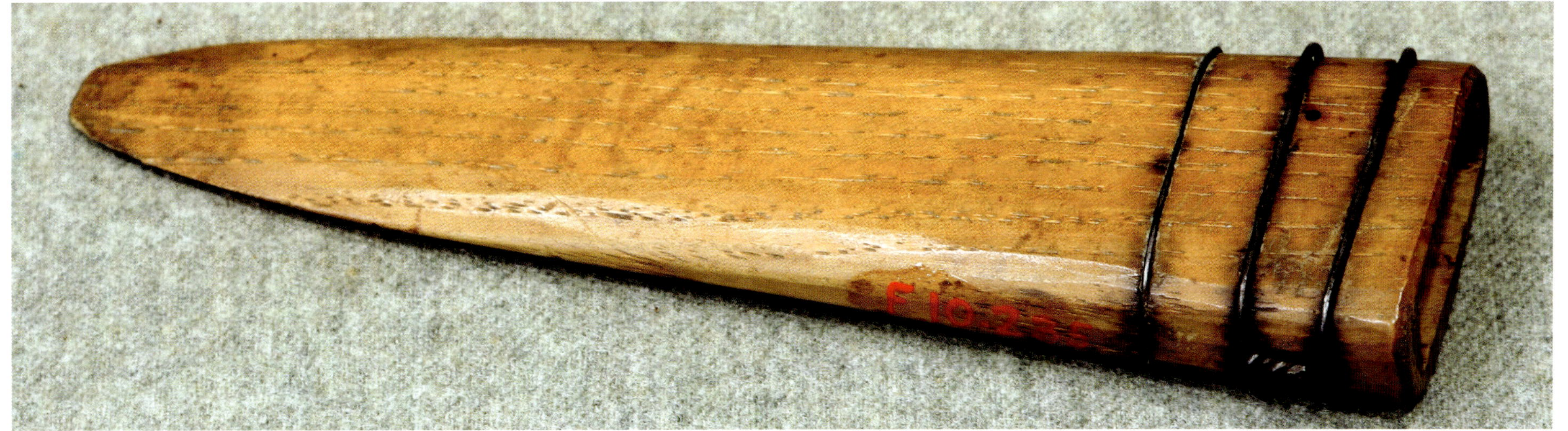

Tobacco Spears. Used to spear plants onto laths for drying.

Tobacco Wagon. Specially designed to carry loaded laths of tobacco to the barn.

Detail showing tobacco spear holders.

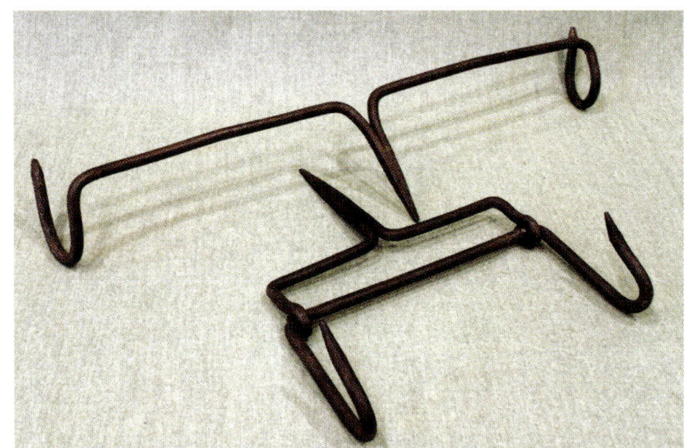

Lifters. Used to lift and hang tobacco.

Lath pullers. Remove tobacco from the laths.

Leaf Sizer.

Sample Tobacco Binder. Binds small quantities of tobacco to analyze its quality.

Cigar Molds.

Cigar Making at Landis Valley.

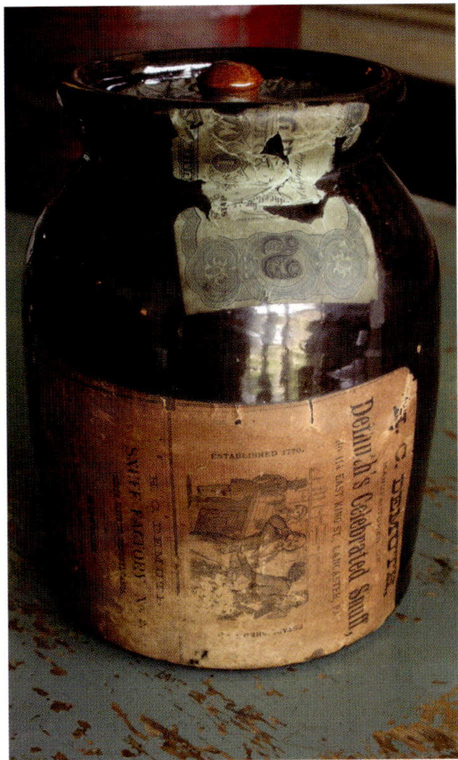

Snuff Jars. Snuff, a ground tobacco product, was often sold in earthenware jars.

Chapter 5

Orchards

When European settlers arrived in England's American colonies, they found few fruit trees. All of our commonly grown orchard trees – apple, pear, plum, cherry and peach, are Old World in origin, but one, the peach, was there to welcome them. The Spanish had introduced the peach into the southwest in the 16th century. Through trade among various Indian tribes, peach pits or stones found their way north. By the time the settlers arrived, peaches were everywhere. Few agricultural pests, and none serious, attacked the fruit. An 1808 description of growing the fruit sounds like paradise to current fruit growers:

> The common mode is to plant young trees grown from the stone without budding or grafting, dropping two stones in a hill of corn, about twenty-five feet apart in squares. They tend the corn in the usual way and the young trees grow with the corn crop at small expense. The knife is never applied to the trees, it being found injurious and to occasion the limbs of pruned trees, heavily loaded, to break. When suffered to grow at pleasure the limbs are flexible and tough and lay on the ground securely when loaded with fruit. They recover their usual position when the fruit is detached. The orchards are generally enclosed to exclude horses and horned cattle. The trees have been known to endure fifty years. They are so easily renewed that it is not uncommon to cut the trees down to the ground occasionally to permit a course of tillage, after which they are renewed by sprouts from the roots.

The transplanted Europeans, the colonists, were soon growing a full range of our familiar fruits, most from seeds with orchards for home use planted on virtually every settlement farm. Apples were preferred: most were seedlings,

The Franklin S. Merkel farm in Berks County, Pennsylvania, had an extensive orchard. It was not unusual for a German family farm to have upwards of two hundred trees to produce fruit for cider making, dried apples or *schnitz,* and apple butter. Note that mown grass surrounded the trees. There was always debate as to whether to use the area for meadow, raising other crops, or to keep it bare earth. Farmer Merkel is seen removing the center branches of the tree so that light can reach fruit in the center. Winter was the favored time for pruning and basic orchard tidying. Not only were the trees dormant at that time, but other farm chores slowed. *Courtesy: Schwenkfelder Library and Heritage Center. H. Winslow Fegley photographer*.

PLANT A PROFITABLE ORCHARD

THE DEMAND FOR GOOD FRUIT is ever increasing at highly profitable prices. Nothing else on the farm brings such great returns per acre. Fruit trees in comparison with other crops require little time and can be grown by anyone who follows the instructions we furnish.

DON'T WASTE TIME with inferior stock at cheap prices! Plant hardy, tested and reliable Pennsylvania grown trees for a permanent and profitable investment. Varieties, prices and full instructions on application.

800 ACRES devoted to growing fruit and ornamental trees, shrubs, vines and plants of the highest quality, healthy, vigorous and productive.

HOOPES, BRO. & THOMAS CO.

PHILADELPHIA OFFICE, STEPHEN GIRARD BUILDING WEST CHESTER, PENNA.

An advertisement in the Chester County Directory [1914] tells all. Typically, the ad features the end product – the harvest. Note the fruit picker's bag and the harvest ladder. *Collection: Mr. and Mrs. Michael B. Emery*.

but very early some were grafted from superior trees. Eating fresh fruit was desirable, but secondary. Most tree fruits were grown to turn into drink, to dry, or to use for animal feed. The most popular of farm families' "Comfortable drinks" was hard apple cider. Pears made perry, and peaches and plums were favored for distilling into brandy-like drinks.

A unique event in the history of orchard culture revolves around the exploits of picaresque John Chapman, A. K. A. "Johnny Apple-Seed," who from 1800 to 1840 spread seedling apple trees from Western Pennsylvania on west to the prairie states. Collecting seed in older farm settlements, he carried them west in his saddlebags and then sowed them in clearings he protected with brush fences. From these nursery beds he harvested the seedlings he gave to frontier farmers.

Commercial fruit growing didn't begin until after the Revolution when the urban population grew large enough to support markets for the produce. Paralleling the move into commercial fruit growing is the emergence of commercial nurseries led by the Prince Nursery of Long Island, N.Y., the

Orchard on the Franklin S. Merkel farm in Berks County, Pennsylvania. *Courtesy: Schwenkfelder Library and Heritage Center. H. Winslow Fegley photographer*.

Landreth nursery in Philadelphia, and that of William Coxe of Burlington, New Jersey. With the rise of nurseries came large-scale production of grafted trees of known varieties rather than seedlings. By 1800 about twenty known varieties of apples were commonly grown, including Esopus Spitzenberg, Tolman Sweet, and White Winter Pearmain. A hundred years later there were hundreds of varieties in cultivation. Fewer varieties of other fruits were known and remarkably many are still commonly grown, including Bartlett and Seckel pears, Ox Heart and Black Heart cherries, and Green Gage plums.

Modern commercial fruit growing is a product of the industrial revolution and a function of the improvement of the transportation and especially the railroad which made widespread distribution of fruit possible. With large scale production came the increase in pests, many introduced along with imported nursery stock and deluxe fruit. The idyllic state of early fruit growing was turned into a war of man versus insects, bacteria, and fungi, many of which were not adequately understood. Power spraying was first introduced in the late 1880s—and did not become common until the early 20th century, with many early fungicides and insecticides being highly toxic to both insects and man.

Industrialism helped the farmer with many new supplies and tools, but it helped the orchard grower less than others. Tractors and power mowers could cut the grass under trees and power sprayers could be employed, but trees still have to be pruned by hand and fruit mostly harvested the same way. While increased use of chemicals have raised yields, they have also led to a marketplace intolerance of any but perfect-looking fruit—something that nature apparently abhors. Increasingly fruit growing became a highly specialized form of farming located in areas where climate, soil, and water conditions were especially suited to orchards. Today small numbers of cared for fruit trees are more likely to be found in suburban back yards than to be seen on rural farmsteads. There is also a movement to provide the market with organically grown fruit, which is more complex to grow successfully than by conventional methods.

Peach trees are usually the first fruit trees to blossom in our temperate climate. This postcard, c. 1940, was published in Asheville, North Carolina. *The Cottage Collection*.

Often stretches of cropland separated orchard areas. A two-horse team working a plow poses between ranks of blossoming apple trees. *Collection: Mr. and Mrs. Michael B. Emery*.

The honey bee was highly prized for both the pollination of orchards and the production of honey. Most traditional man made hives or bee skeps were bell-shaped and made of rye straw. As the 19th century progressed many "how to" manuals were published to help the novice bee keeper. *Book Cover, Collection: Mr. and Mrs. Michael B. Emery*.

How I Produce
COMB HONEY
BY
Geo. E. Hilton,
FREMONT, MICHIGAN.
PRICE - - - 5 CENTS.
Second Thousand.
"GLEANINGS" PRINT, MEDINA, O.

Bee keepers pose in front of their hives. Some are homemade, others, like those with the gabled tops, were commercially manufactured. The hives are sheltered under a lean-to. Unlike the modern age, hives were generally not moved but stayed on the farm. Different seasons produced different flavored honey. Some people preferred apple blossom honey; others favored clover or buckwheat. *Private collection.*

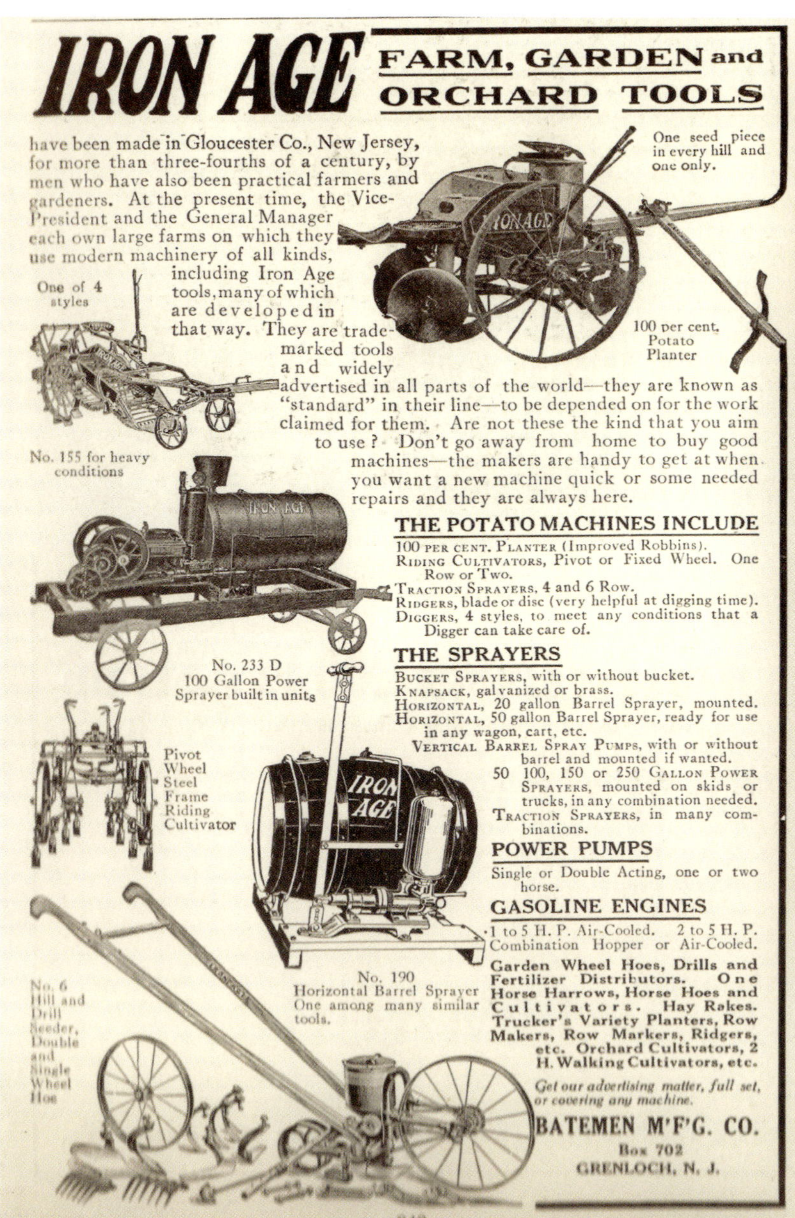

By the late 19th century insect and fungal pests made spraying a necessity. Power sprayers were widely advertised as in the Chester County Directory of 1914. *Collection: Mr. and Mrs. Michael B. Emery*.

Fruit harvesting is hard work, done mostly by hand. Peach trees are low-growing. Apple picking often required ladders. Note the peach baskets and the apple crates. *The Cottage Collection*.

A woman drying "*schnitz*": apples cut into eighths. Sweet dried apples, unpeeled, were used in cooking and making *Hutzelbrieh*, stewed dried fruit. Sour apples were peeled, dried, and used in pies where the combination of sweet and sour flavors was highly prized. The trays used for drying these apples were specially made to fit the outdoors baking oven used for fruit-drying after the household baking was done. *Schwenkfelder Library and Heritage Center. H. Winslow Fegley photographer*.

Many apple varieties were prized for their keeping quality. Apples were often stored in earth or root cellars, before use or sale.

Cider making is a simple process. First the apples are ground, using a grinder (seen to the left in the woodcut illustration.) The pulp is then pressed to extract the juice, or cider, which could be drunk sweet or, most likely, fermented into "hard cider" a very common drink. The splendid large cider press was collected by the Landis brothers but, because of its size, it couldn't be properly preserved. Photographed by Henry K. Landis in the 1940s, due to weathering only the huge wooden screw, now safely housed, remains.

"Sippin' Cider Through a Straw" was a popular song written in 1919, but its message is an early one. Connoisseurs sample cider using pieces of rye straw, probably in front of a mill in Berks County, Pennsylvania. The grinder and press, in one, is a combination of wooden and iron parts. Photographer H. Winslow Fegley (1874-1944) has left us with wonderful images of rural Pennsylvania with many of his images collected in a book, *Farming, Always Farming*. His contemporary Henry K. Landis photographed a little boy at the cider barrels, an important farm utensil. Note the "bung" or faucet. To the left, a copper apple butter kettle rests on an iron stand. Sliced apples and cider were the chief components of the enduring delicious spread. *H. Winslow Fegley photographer. Schwenkfelder Library and Heritage Center*.

LIGHTNING PEACH PARER.

PAT. AUG 17 1869

A continuous and urgent inquiry for a machine for **Paring Peaches**, has been ringing in our ears from all Peach-growing sections, for the past five years, and in response to this universal appeal, the manufacturers of the **Lightning** and **Turn-Table Apple Parers** have at last succeeded in obtaining and securing a device for **Holding** and **Paring Peaches**, which is as practical and economical as the Apple Parer, and cannot fail to come into immediate and general use. They pare **Apples** also, as well as any **Apple Parer.**

They were exhibited in the Fall of 1869, at the NEW YORK, OHIO, INDIANA, ILLINOIS, WISCONSIN, MINNESOTA, and ST. LOUIS FAIRS, and in every instance the judges were delighted with the rapidity and perfection of their work, and awarded them the highest Premium. A limited quantity of them has been made for distribution this year, in anticipation of the immense sale that must speedily follow. On receipt of $1.50 we will send a sample machine by Express.

SARGENT & CO., Sole Agents,
70 Beekman Street, New York.

P. S.—We are also Sole Agents for the **Lightning** and **Turn-Table Apple Parers.**

Dried Peaches.

A few bushels of dried peaches of the best quality. W. P. SHARPLESS.

Commercial quantities of fruit were often preserved in a special dry house, or *Daerrhaus*. After the fruits were pared and/or peeled, slices were laid in special screen-bottomed drawers that flanked a stove. Now relocated at the Landis Valley Museum, this dry house was built in Landisville, Pennsylvania. The advertisement for dried peaches is mid-19th century,

From the Landis Valley Collection

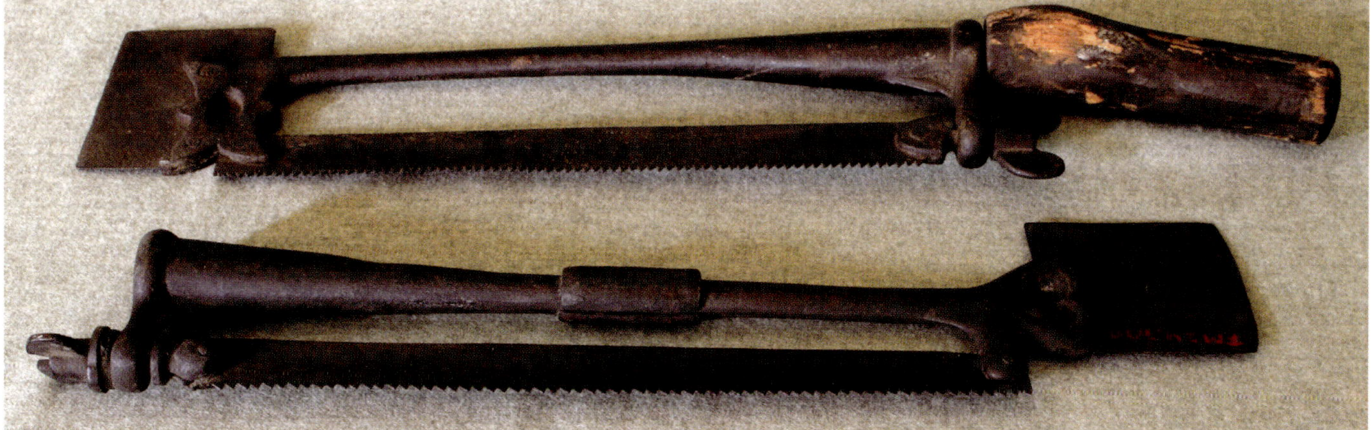

Pruning Saws.

Pruning Hooks.

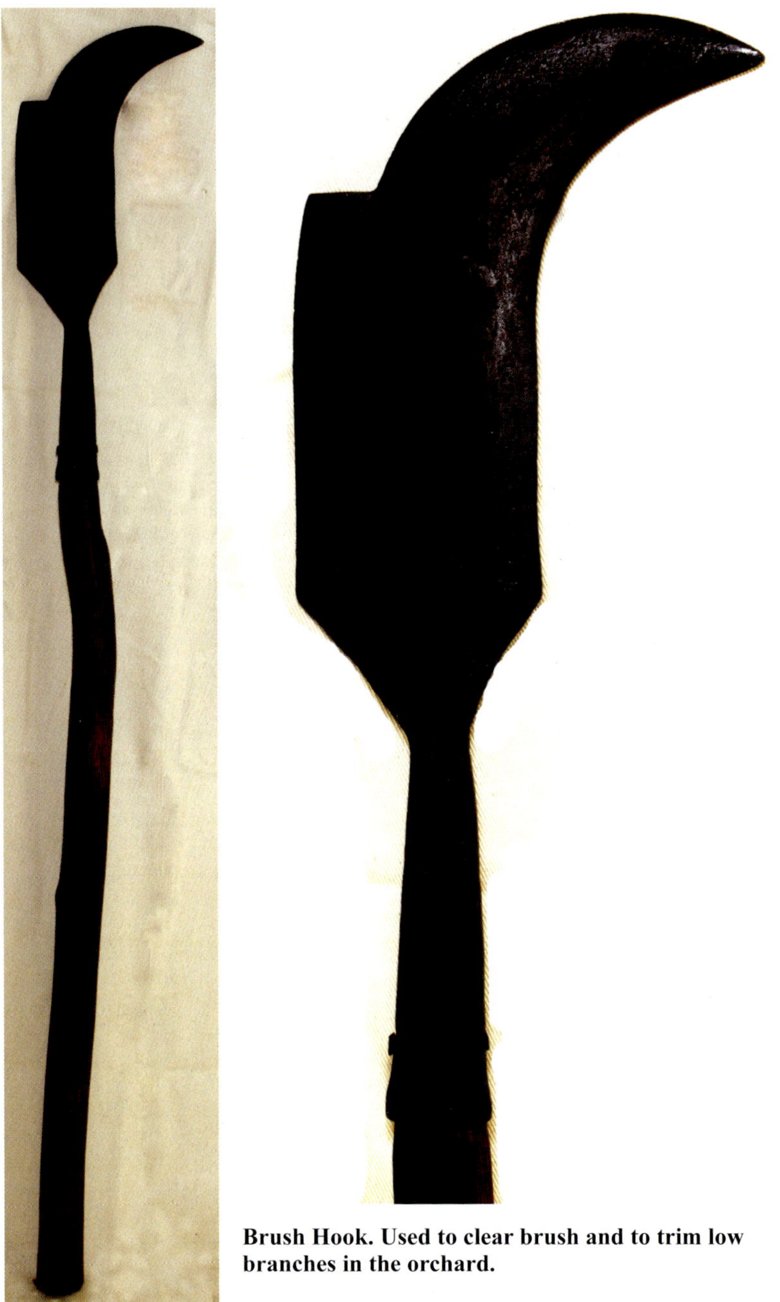

Brush Hook. Used to clear brush and to trim low branches in the orchard.

Grafting Knives.

Rye Straw Bee Skeps or Bee Hives. Traditional forms that date to antiquity.

Bee Boxes or Bee Hives. Commercially produced "modern improvements."

Bee Keeping Demonstrations at Landis Valley.

Hive Smokers. Used to anesthetize bees so honey combs can be removed from the hives.

Orchard Ladder.

Split Oak Peach Basket.

Split Oak Apple Basket.

Heirloom Apples at Landis Valley.

Cherry Pitter.

Drying Rack. Used in a bake oven to dry fruits and vegetables.

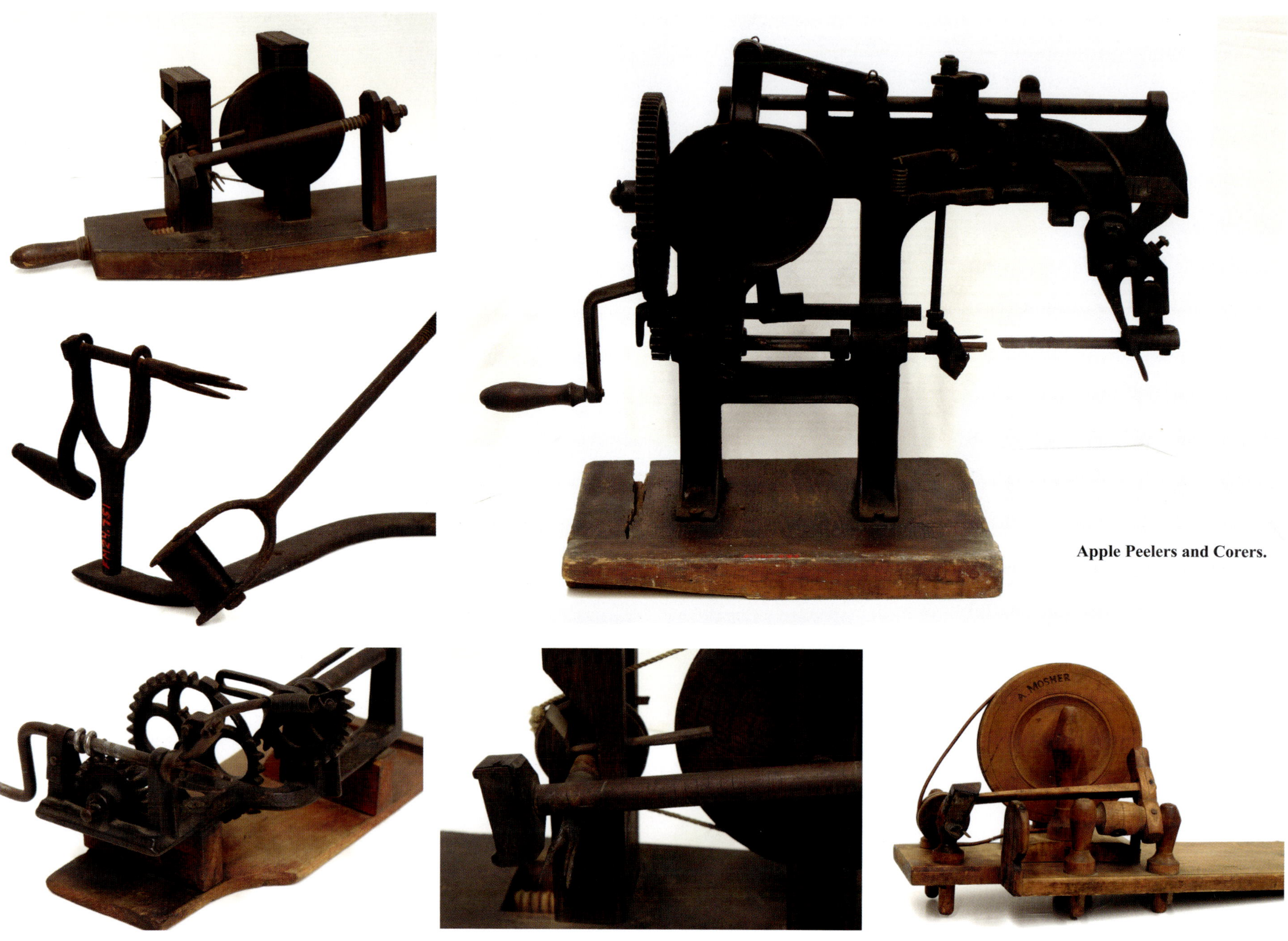

Apple Peelers and Corers.

Apple Butter Making at Landis Valley. Peeled, cored apples are cooked down in cider with spices and sugar into a thick spread.

Apple Butter Paddle.

Copper Apple Butter Kettle.

Apple Grinder and Cider Press.

Schnitz Basket. Dried apples or *Schnitz* was a staple in the Pennsylvania German diet. Typically it was stored in lidded rye straw baskets.

Mason Jar. Used for canning apple butter as well as fruits and vegetables.

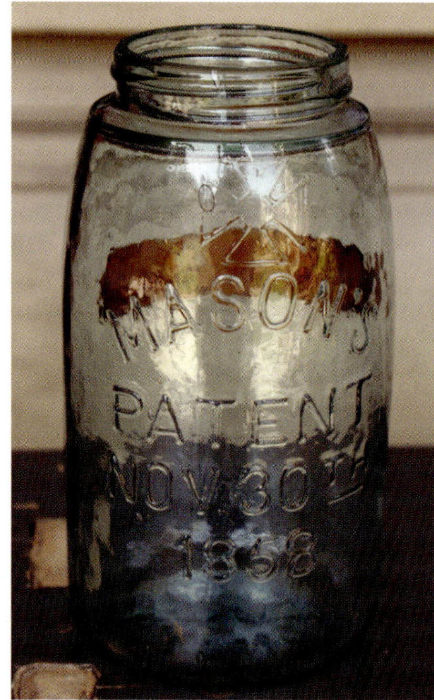

Commercial Apple Butter Crock.

Stoneware Apple Cider Pitcher. Hard cider was the most common table drink in the 18th and most of the 19th century among many rural people.

Stoneware Cider Jugs. Hard cider was often carried to the fields to slake thirst.

Copper Still. Many farms had stills, which were used to make liquor from grains or fruits.

Chapter 6

Market Gardens and Poultry

From colonial times into the early modern period table gardening and the keeping of poultry was considered to be woman's work. Traditionally vegetables were a minor part of the diet and tending of chickens was a proud source of "egg money" for the farm wife. Aristocrat William Penn could be proud of the variety of vegetables he had grown for his table at his Pennsbury Manor estate bordering the Delaware River in 1685.

> Kidney Beans and English Pease of several kinds, with English Roots, Turnapes, Carrets, Onions, Leeks, Radishes and Cabbidges. … Also I have such plenty of Pumpkins, Musmellons, Water Mellons, Sqwashes, Coshaws, Buck-hen, Cowcumbers and Sinnels of Divers kinds.

But most families had a more limited range centering on cabbage, turnips, onions and perhaps potatoes. While more species were added as the years went on vegetables were not recognized as an important, let alone a necessary, component of the diet until the late 19th century – and even then old habits died hard. Since gardens were small, tools were simple and the forms of most garden tools – rakes, spades, hoes and the like would remain stable and their forms essentially unchanged over the years – and, indeed, still easily identified by contemporary home gardeners. Before 1840 virtually every family except some living in larger cities had their own garden growing the traditional limited variety. A revolutionary addition to home gardens was the tomato, which became common after about 1875. Sweet corn, another now essential home garden crop, had already entered the garden about 50 years earlier.

The growth of urban population centers resulting from the industrial revolution led to the rise of market gardening with men cultivating large plots, or truck gardens, dedicated to serving the marketplace. From about 1850 to 1875 market gardening around cities like Philadelphia was exceptionally profitable because only these nearby sites could supply the demands of the new urbanized society. Especially treasured were the earliest crops. Those who could bring in the first radishes or lettuces to the market could command

A scarecrow stands guard over a Pennsylvania truck patch early in the 20th century.

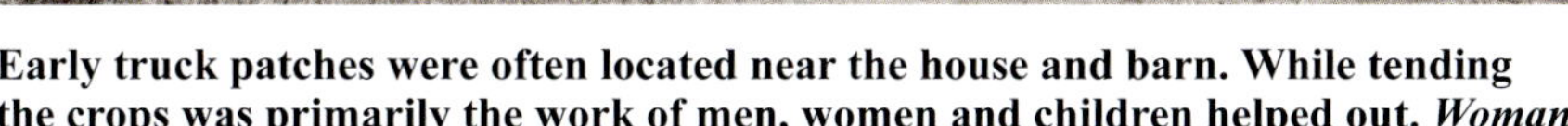

Early truck patches were often located near the house and barn. While tending the crops was primarily the work of men, women and children helped out. *Woman and boy in garden,* *Collection: Mr. and Mrs. Michael B. Emery. Painting: The Cottage Collection.*

a premium price. Many farmers grew crops in hot beds and cold frames to get early produce to market.

Markets for vegetables changed radically after 1890 when the development of refrigerated railroad cars made it possible and profitable to bring common vegetables to market from distant places like Florida and California. Most commonplace produce would soon be found almost every day: lettuce in January, tomatoes in April, and string beans in May.

While the year round production of fresh table vegetables headed south and west, the rise of the canning industry and later frozen foods led to the growth of large scale production of seasonal produce in the East and Midwest, often on contract for food processors. While traditional vegetable growing and market gardening was hard slow work, after 1900 there was a revolution in technology with the development of machinery that could do many of the major operations – some planting and harvesting, all cultivating, fertilizing and spraying. Hybridizers provided vegetable varieties that all ripened simultaneously. Until about 1930 most large scale commercial gardeners believed that machines couldn't take the place of horses in such precise operations as seeding and close-row cultivation. It was also feared that tractors would compact the soil. The development of lighter, smaller tractors equipped with rubber tires changed that picture, and after 1935 a whole range of specialized gardening attachments were developed, including multiple-row seeders, side dressers, and plant setters. A revolution had taken place and most farms that now raised vegetables did so on a large scale for canners or freezers. On many farms the housewife's garden disappeared and, especially after World War II, farms became completely specialized. Farm wives would most likely shop for table produce at their local store.

In the late 20th century, a new movement appreciating local flavors and the perceived energy efficiencies of eating locally grown produce led to the development of a new variety of vegetable farms where the small family farm or the Community Supported Agriculture (C S A) enterprise concentrates on quality specialty crops often sold directly to the consumer. While most of these operations use modern machinery, most complement their machines with the use of hand tools their agrarian forebears would recognize from hundreds of years before --those similar to the ones used by generations of housewives. Indeed, increasing numbers of specialty farmers are women.

A similar cyclical tale centers on poultry growing. It wasn't until about 1900 that poultry husbandry was considered a proper occupation for a man. For the most part, "Poultry keeping," according to an agricultural writer in

1875, was "regarded by the masses as an insignificant business adapted to, and conducted by, the women-folk and children."

Chickens have a long history in America having been introduced into the West Indies by Columbus in 1492 and by the English into Jamestown in 1607. Because they were easy to transport and feed, chickens were common on transatlantic trips made by early explorers and settlers. That said, until about 1840 the birds tended to be of minor concern to farmers' families and they were often expected to literally scratch for their livelihood. Almost every farmer had "dung hill fowls" for family use. An important source of feed for these birds of often unknown lineage was undigested grain and seeds that had first passed through other farmyard livestock. Hens gave few eggs – maybe sixty or so a year (as opposed to the 300 expected of modern producers), mostly during spring and summer, which were laid in nests that had to be first discovered and then carefully watched. Eggs for year-round consumption had to be preserved – usually in crocks in a strong solution of water glass (sodium silicate). Chicken flesh was prized, but laying hens were never slaughtered and fowl for meat was scarce. Many cultures have a variation of the observation that, "When a poor man eats a chicken, either he or the chicken is sick."

Also in the farmyard were a small flock of geese whose feathers were harvested for use in bedding, a few domesticated turkeys and perhaps several ducks. Duck eggs were especially prized for the richness of their yolks. Housewives would generally trade their excess chickens and eggs with neighbors or they might sell them at a neighborhood store, which in time would sell the eggs and live poultry to peddlers who sold them in nearby towns and cities. Enterprising farm women would also sell eggs directly in nearby urban areas, if any. Interestingly, the first semi-large commercial transport of poultry from countryside to cities occurred in the fall when large flocks of domesticated turkeys and geese were herded into cities. Julius F. Sachse recorded an early 19th century trip from Lancaster to Philadelphia.

> This was not an easy task. They were apt to crowd together and trample each other to death. To prevent this they were divided into lots of 50-75 with a 'shooer' in charge of each lot. He had a long pole with a piece of red flannel fastened to the end. The best time made on the road was not much over a mile an hour. As soon as it commenced to grow dark the fun began. The birds were determined to go to roost and notwithstanding all the efforts of the drivers they generally did. The stampede usually took place in passing an orchard or grove of trees. In much less time than it takes to tell it the trees were black with birds and the day's journey was ended for the turkeys. Not so for the drivers, who had to watch the birds all night to prevent theft.

By early in the 20th century large scale truck gardening was common throughout America. Celery growing was a specialty. The horses draw a piece of equipment designed to mound earth around the growing plants to bleach the stalks. ***The Cottage Collection.***

The legend on the reverse of a postcard mailed in 1946 describes truck farming, mostly driven by the needs of canneries and freezers: "BEAN PICKING IN RURAL NEW YORK – The Empire State ranks first in the growing of snap beans in the nation, as well as green peas, onions and cabbage. New York State leads many Central States in agricultural pursuits largely because of the fine soil, grazing lands and abundant rains. Truly New York is the 'State that Has Everything.'" *The Cottage Collection.*

While there would be great advances in the raising of poultry after 1900, the role of the farm wife in raising eggs and poultry would remain important well into the 20th century.

Even though there were some commercial poultry operations near large cities as early as the 1850s they, of course, were miniscule by modern standards. Both authors, while of different generations, 36 and 72 years of age, remember buying eggs right from farm families.

For the most part, farm women well into the 20th century reared their chicks under broody hens, fed them homegrown corn, and housed them in small chicken houses, with many having yards enclosed with chicken wire. Early in the 20th century also saw the introduction of many chick-related tools and utensils including incubators, which facilitated the hatching of chicks. Since 1900 purebred birds have largely replaced the mongrel birds of historic usage. The White Leghorn became the preferred egg-laying fowl from 1900-1940 and the Rhode Island Red and the New Hampshire Red became favored for meat.

By and large, from the end of World War II we have witnessed the large scale commercialization of poultry raising. Since the 1970s, at the latest, the factory farm has been the rule in chicken production with individual farms housing hundreds of thousands of birds at a time and eggs eventually being sold increasingly by brand, as are chickens. Today fewer people buy generic dressed chickens than buy Purdue, Tyson, or Empire Kosher. Even more specialized than chicken farming has become, turkey husbandry that has developed into a subspecialty of its own. Just as modern methods make it possible to have inexpensive eggs and chickens available year round, now fresh turkeys can also be made available. Specialty farmers also raise geese and ducks, and a wide variety of game birds, including quail and pheasants.

Mirroring a movement toward locally grown and organic vegetables, is a growing market demand for birds and eggs free of hormones, antibiotics, and insecticides and more popularly "free range," *i.e.* chickens that are not caged. Very few, however, advocate the return to the dung fowl. As with vegetables we have been witnessing the growth of a boutique poultry industry that produces birds by variations of older techniques, now even using antique or heritage varieties of the fowl. There is, however, as yet no push to use broody hens; virtually all fowl raised in America is hatched by specialists operating commercial hatcheries, some specializing in old varieties with colorful names like Buff Orpingtons, Top Knots, or Bearded Silver Poland. All of this is very specialized, but it fits within the realm of contemporary mores.

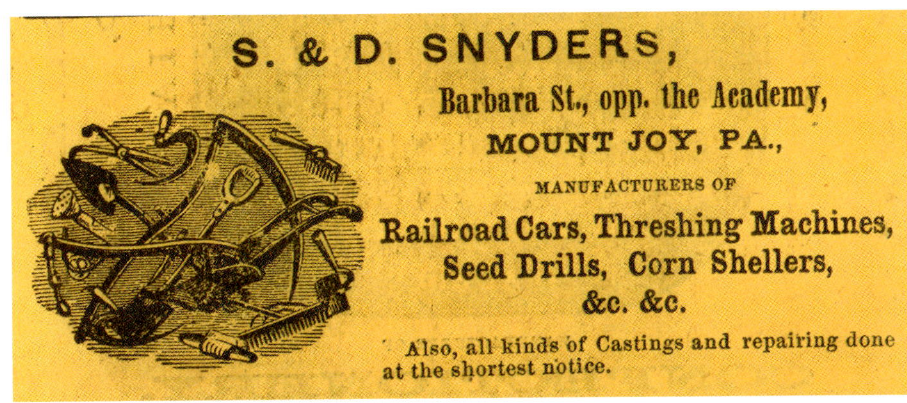

Tools, most of which were used in gardens, are shown in an advertisement for "S. & D. Snyders" of Mount Joy, Pennsylvania, although the copy offers more elaborate equipment such as "Railroad Cars" and "Threshing Machines." Spades, rakes, hoes and sprinkler heads are all garden staples.

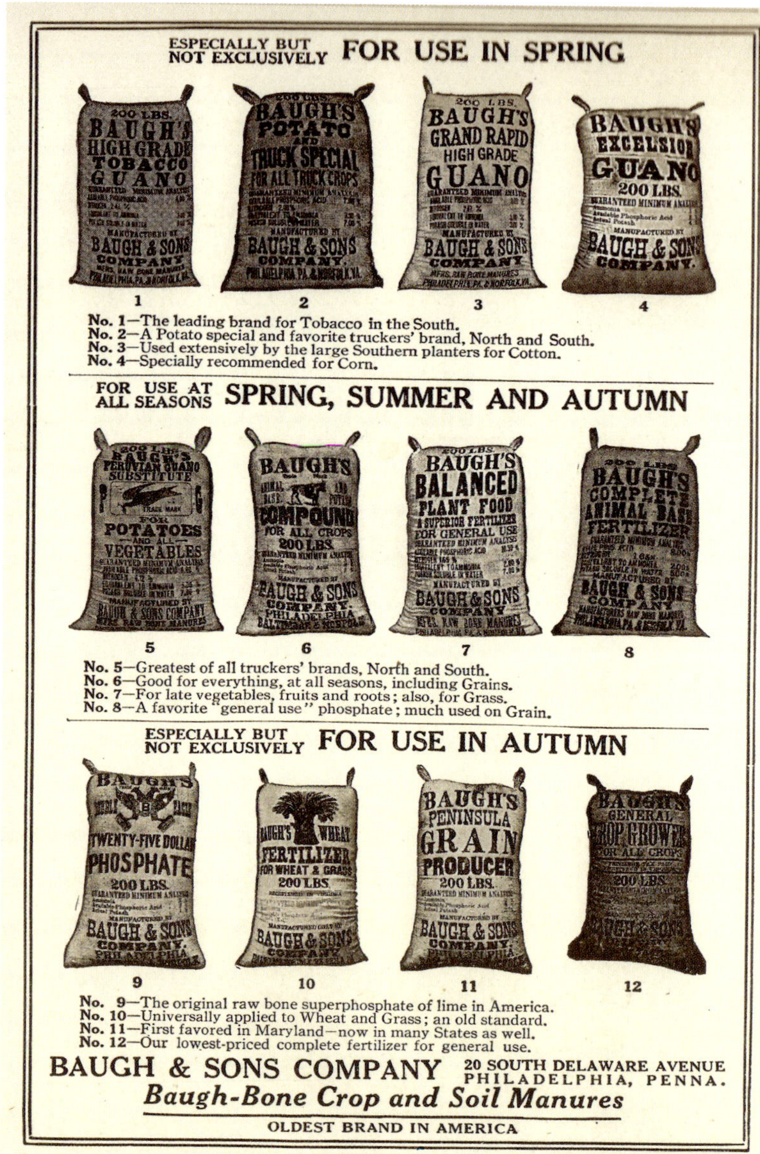

Fertilizers are important in the farm scene. After the Civil War, increasingly farmers would substitute or supplement manure with commercially available fertilizers. In the advertisement for Baugh and Company note that, "No. 2" is "A … favorite truckers' brand" and that "No. 5" is the "Greatest of all truckers' brands, North and South." *Collection: Mr. and Mrs. Michael B. Emery.*

South Shed
Used for Light work = Made to Rese
Conestoga Wag

Boys Work Wagons
Later Replaced by Wheelbarrow
South Shed

A page from Henry K. Landis' tools scrap book shows light work wagons. Note that one would "later [be] Replaced by Wheelbarrow".

A "hired girl, probably a farm wife earning extra money," pushes a wheelbarrow near the kitchen garden at Landis Valley in the early spring, around the turn of the 20th century. Rural women usually shaded themselves when they went outside. Suntans were not considered to be attractive. *Henry K. Landis, photographer*.

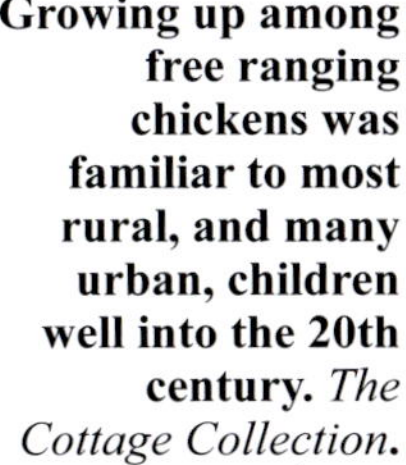

Growing up among free ranging chickens was familiar to most rural, and many urban, children well into the 20th century. *The Cottage Collection*.

Tending the chickens was usually women's work but, on occasion, men would also feed the flock. A variety of pails served to carry the chicken feed. Flocks generally ranged free, but were encouraged to return to roost in chicken houses, which were often locked at night against predators and thieves. *Collection: Mr. and Mrs. Michael B. Emery*.

Emma Caroline Landis (1842-1929), the matriarch of the Landis family, poses on the back porch of her home with her egg basket in hand. Emma was always proud of her flock of chickens and enjoyed earning her "egg money." The photograph is by her son Henry.

Homer H. Hewitt of Williamsport, Pennsylvania, featured "High Class Poultry" for farmers and urban householders. Among his offerings were, the no-longer-seen-today, Partridge Cochins and the very common White Leghorn, which, in its improved form, is the most common laying hen employed in today's large-scale egg operations. *Collection: Mr. and Mrs. Michael B. Emery.*

B. H. Greider of Rheems, Pennsylvania, promised that his catalog was the route to doubling your poultry-raising profits.

Duck farming was developed as a specialty in various locales. In the late 19th and early 20th centuries the area around Allentown, Pennsylvania, and much of eastern Long Island in New York State were famed for their duckling production. Some Allentown duck farms welcomed visitors and became early tourist attractions. The duck farm illustrated evolved into contemporary Dorney Park, a famed regional amusement park. *The Cottage Collection*.

Most farms had a few turkeys and, like ducks, their culture became more and more specialized. The small turkey flock on the turn of the century postcard bearing the surprising legend "Merry Christmas" was typical of the small pre-specialty farm era. A turkey is part of the bounty of the American farm celebrated on a Thanksgiving Card. *Merry Christmas, Collection: Mr. and Mrs. Michael B. Emery, Bounty Card: The Cottage Collection*.

JELLIFFE, WRIGHT, HOAG & CO.,
Wholesale Commission Dealers in
COUNTRY PRODUCE,
13, 15 & 17 Centre Row, and
16 and 18 Merchants' Row,
West Washington Market, NEW YORK.

....................Packages.......................Lbs.

FROM

Philadelphia,..............................191

M..............................

Bought of JAMES B. SHAW
Dealer in
Prime
Country Produce
HONEY BROOK, PA.
Stalls 957
Reading Terminal Market

The market houses found throughout Pennsylvania and many other states were important outlets for both eggs and the product of the truck patch. Henry H. Landis and a partner rented a stand in the City of Lancaster's Northern Market for 1875. Most farmers, however, sold their crops through middlemen such as James B. Shaw, who dealt in "Prime Country Produce" at Philadelphia's Reading Terminal Market, or commission dealers, such as Jelliffe, Wright, Hoag & Co. "of New York City."

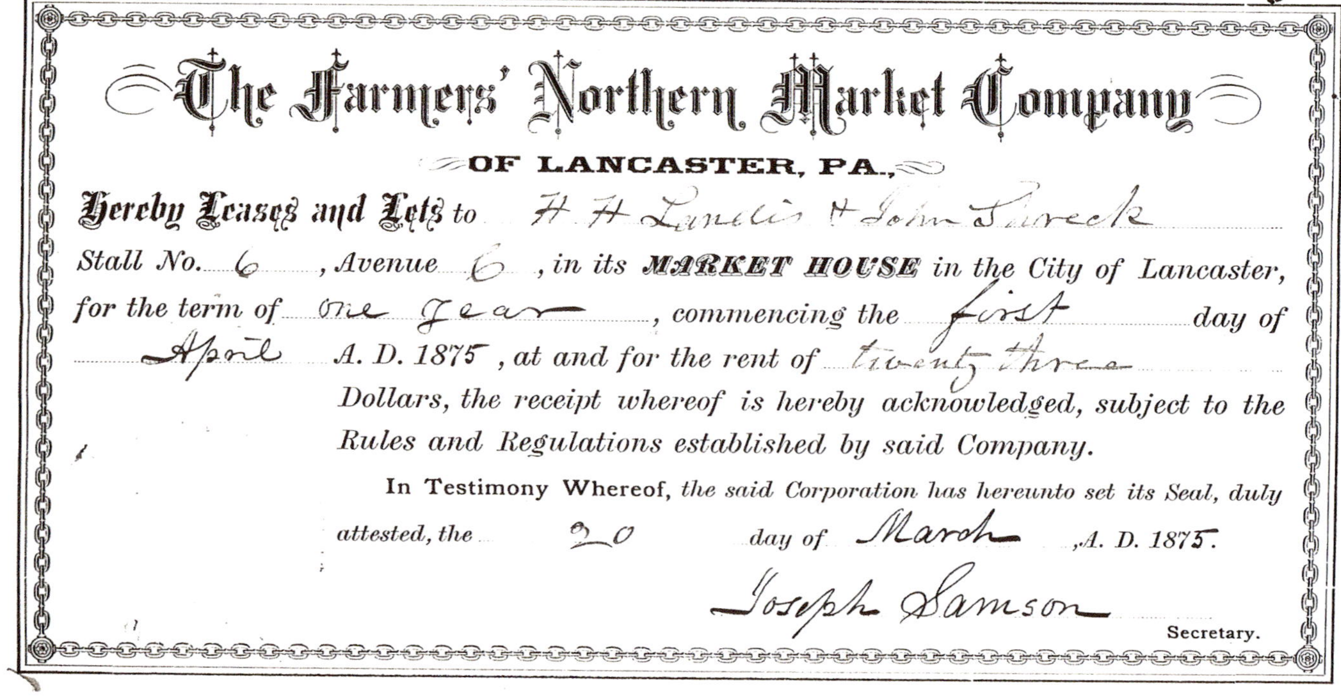

The Farmers' Northern Market Company
OF LANCASTER, PA.,

Hereby Leases and Lets to H. H. Landis & John Sureck

Stall No. 6, Avenue 6, in its MARKET HOUSE in the City of Lancaster, for the term of One Year, commencing the first day of April A. D. 1875, at and for the rent of twenty three Dollars, the receipt whereof is hereby acknowledged, subject to the Rules and Regulations established by said Company.

In Testimony Whereof, the said Corporation has hereunto set its Seal, duly attested, the 20 day of March, A. D. 1875.

Joseph Samson
Secretary.

From the Landis Valley Collection

Horse Drawn Shovel Plow.

Burdock Weeder.
Used for extracting deep-rooted weeds.

Garden Spades.

Spading Fork.

Garden Hoes.

Garden Hoe and Cultivator Heads.

Hand Cultivators.

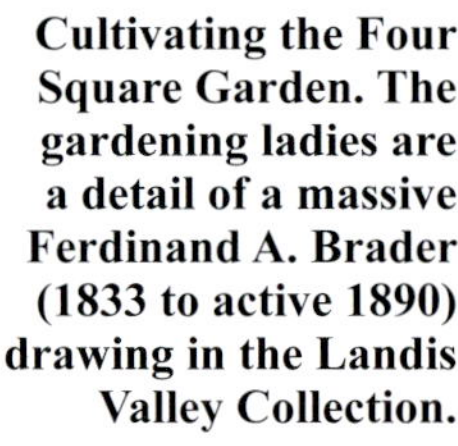

Cultivating the Four Square Garden. The gardening ladies are a detail of a massive Ferdinand A. Brader (1833 to active 1890) drawing in the Landis Valley Collection.

Four Square Garden at Landis Valley.

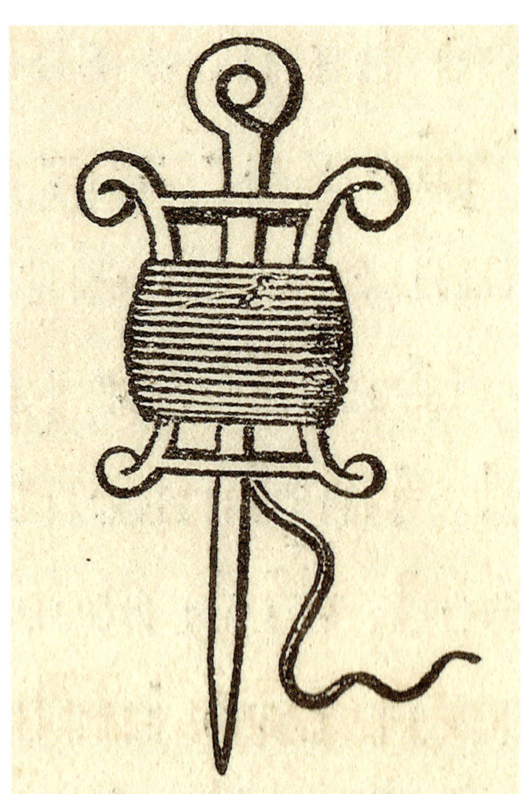

Straight Line. Used to lay out straight rows in the garden.

Garden Cuffs. Used to protect a woman's garment during gardening.

Dibbles. Used in planting seeds and transplanting small plants. Made in wood and metal.

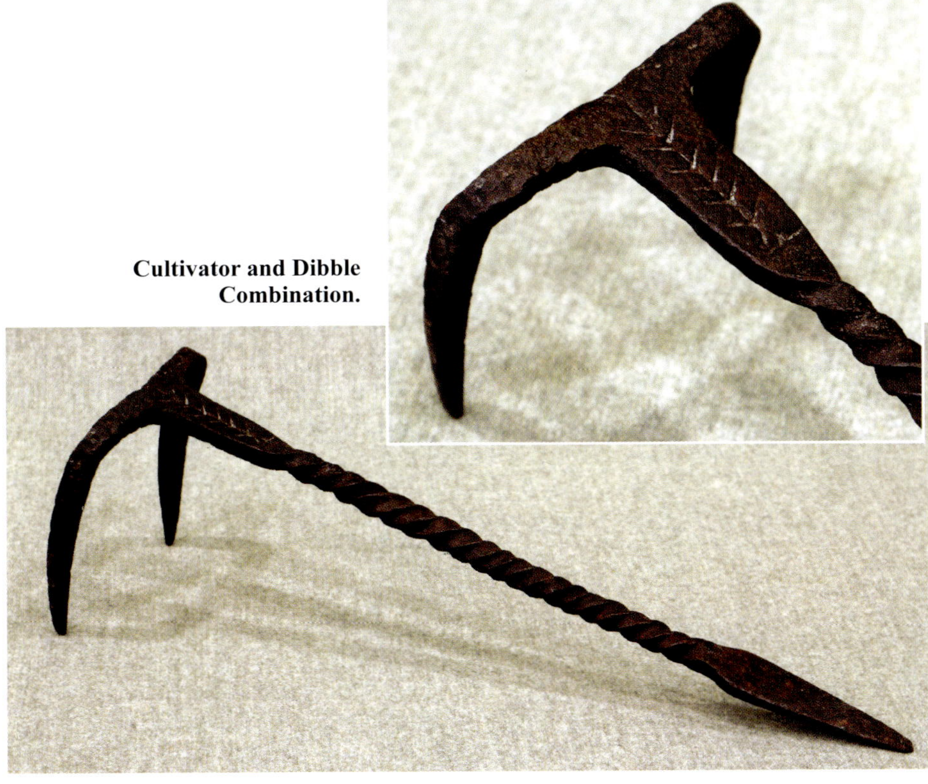

Cultivator and Dibble Combination.

Garden Cultivator.

Garden Harrow/Cultivator. A rare, small hand-powered implement.

Noise Makers. Used to scare birds from gardens. The chore was often performed by children.

Garden Wagon.

Garden Cart.

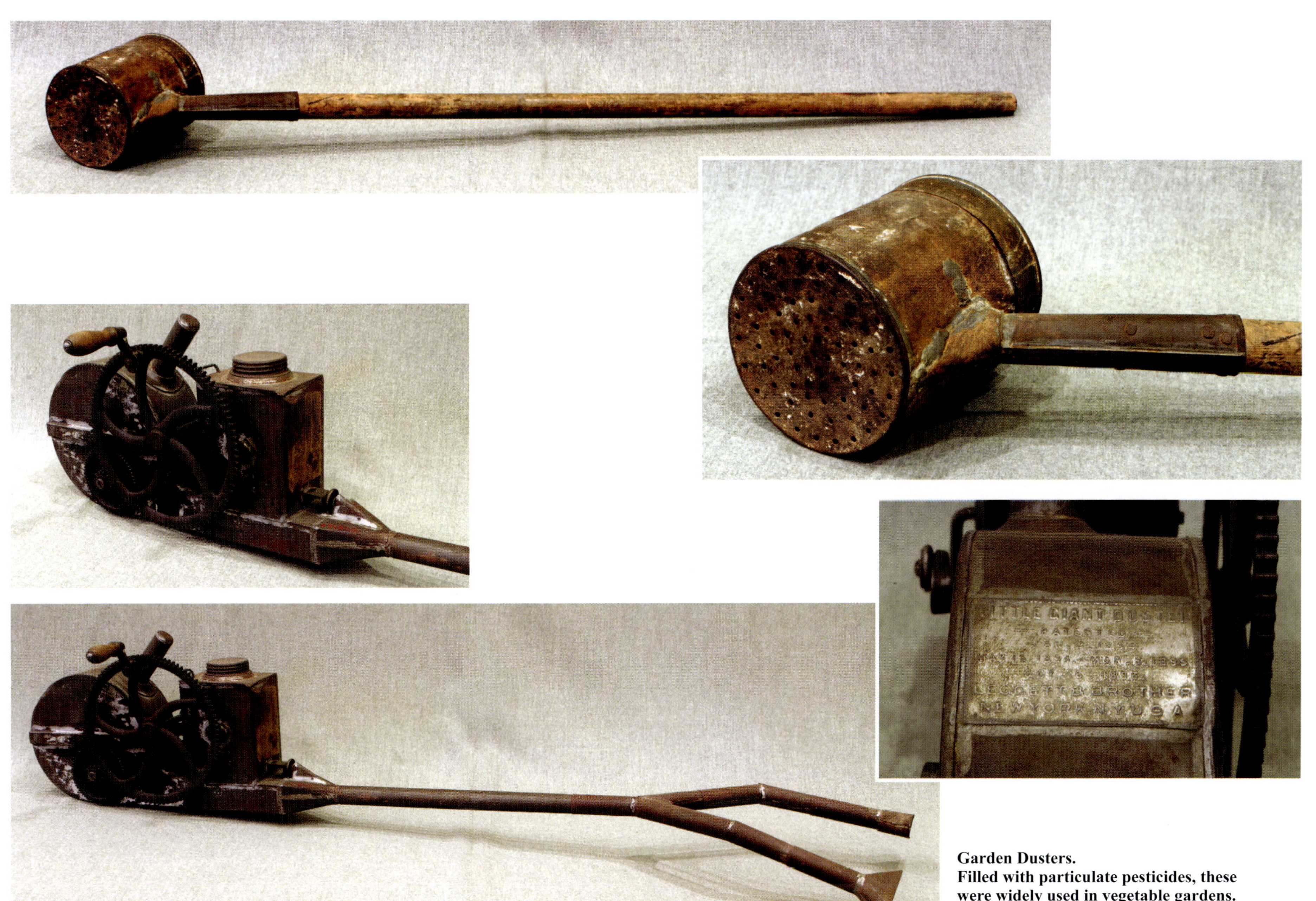

Garden Dusters.
Filled with particulate pesticides, these were widely used in vegetable gardens.

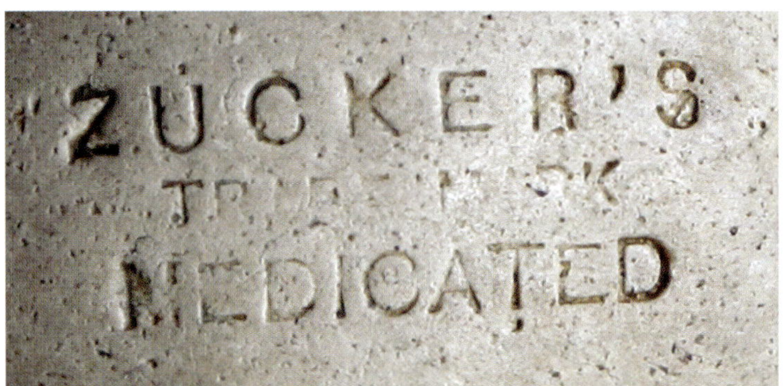

Zucker's Medicated Egg. Filled with medication, these served as nest eggs and to suppress mites.

Ceramic Nest Egg. Made in Germany.

Nest Eggs. Glass and pottery eggs were placed in chickens' nests to encourage egg laying.

Price Board. Used in country stores to show current prices on local staples.

Eggs from Heirloom Chickens at Landis Valley.

Egg Baskets. These needed to be well-balanced and to sit flat. Usually made of split oak.

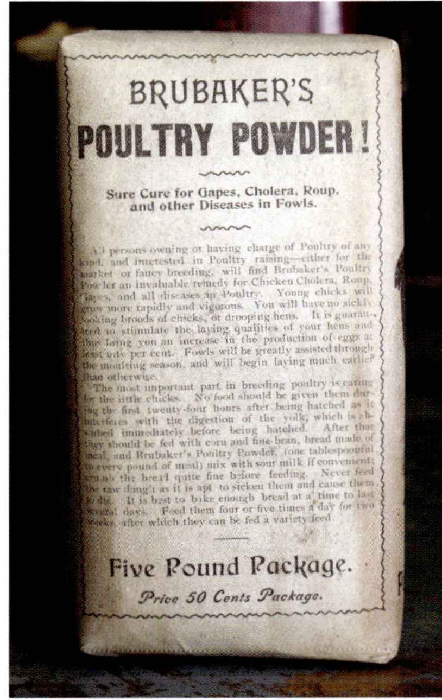

Poultry Powder. Dusted on chickens, this killed lice and mites.

Egg Carrier.

Heirloom Poultry at Landis Valley.

Chapter 7

Dairy, Cattle, Pigs, and Sheep

The importance of food source animals to farmers has changed over the years. Where once dairy was minor, it would become dominant in much of the northeast. Where sheep were once important, they have become trivial. Pigs have remained an important presence.

Every early farm would have a cow or two and several pigs, but the major protein source appears to have been game; as game became exhausted more domestic animals were raised. Traditionally cattle and pigs were treated poorly and were expected to scrounge for their living through most of the year. Both animal types were often left free to roam the woods to browse on vegetation and pigs often became feral. Running wild, they often reverted to the wild boars of their ancestors. Even in winter, with the exception of Pennsylvania German and Holland Dutch farms, cattle were not housed during the winter months. Cold, often undernourished early cattle used interchangeably for milk and meat were poor providers of both. It wasn't until the late 18th century that farmers in general began to realize that feeding cattle and keeping them sheltered in winter would be of great benefit.

Another problem facing farms was that common cattle were mixed breeds, which were very varied in performance. Most were poor milkers and uneven weight gainers (important for meat production.) The first pure-bred cattle to be introduced into America was the Devon, which had been developed in England on the estate of the Duke of Bedford. Like most cattle of the period they were a "triple purpose breed" in that they (oxen) could be used as draft animals, they produced good quality beef (although they were relatively small), and they yielded small quantities of milk high in butterfat content. The Devon were followed by the Durhams or Shorthorns. Most were imported by gentlemen farmers seeking to improve the quality of their herds. Similarly, by the first quarter of the 19th century most of the imported breeds of cattle, familiar today – Guernseys, Holsteins, Herefords, and the Aberdeen Angus – were here.

Beef farming became economically important by the beginning of the 19th century. In Pennsylvania large numbers of beeves were raised in the west and walked by the thousands to the east, by drovers, where they would be grass-fed on the rich eastern grasslands before being slaughtered. Chester

"FIVE QUEENS" reads the reverse of a postcard mailed in 1967. "These five cows represent the Brown Swiss, Jersey, Holstein, Ayrshire and Guernsey breeds. The beauty of each is an inspiration to every breeder of dairy cattle." The "Prize Ayrshire Bull" was illustrated in an 1878 publication. Note the ring in his nose. It allowed a handler to more easily persuade the powerful animal to move. *Queens: The Cottage Collection.*

County became especially known for its high quality pasturage and a little later Lancaster County became the center for raising stall-enclosed, grain-fed animals. These enterprises were all extremely profitable before the railroad, which changed the dynamics of the market. Western beef came to dominate. However, even today small numbers of cattle continue to be "finished" or fattened on eastern farms, mostly in pastures.

Before the day of large-scale feedlots, manure was looked upon as a huge asset and prized by farmers for fertilizing their fields. Produced in quantities that could easily be incorporated into other aspects of agricultural practices, it was recycled to increase soil fertility.

Milk was seldom consumed as a fresh beverage and it was, as it remains, very perishable. Commercially it was only available from farmers whose spreads were in or around towns and cities. In New York, for example, a common sight was a man carrying two large kettles from a yoke on his shoulders trudged through the streets and called out, "Here's the milk, Ho!" Some farmers near the city drove around in carts which were mostly covered and sold milk for six to ten cents a quart.

As cities grew, however, so did the demand for fresh milk, but until the late years of the 19th century most milk was turned into butter or cheese with butter making, an often arduous task being relegated to the farm wife. An article in an 1840 edition of the *Cultivator* describes the process:

> The milk is strained in pans or oaken tubs, holding two pails full. Everything is done in the cellar. The milk is not meddled with until it has coagulated, when each day's or each half day's milk is put in the churn with an equal quantity of water, cold in summer warm in winter, to bring it up to the proper temperature, which is 55 to 60 degrees. The churn is made in barrel form, of oak, hoped with iron. For six or ten cows the churn should hold 30 galons. They rarely exceed two barrels, as in large dairies they prefer to churn several times a day. Churning is never done by hand except for a single cow. In small dairies the churn is worked by a dog or sheep, the latter being preferred; the larger dairies have water or horse power.

Many farm wives took pride in the quality of their butter and enjoyed the proceeds of their labor. "Butter money," like "egg money," was usually considered the farm wife's discretionary income.

Cheese was also farm made in varied quality by individual farm wives. The Philadelphia area became famed for a soft spreadable cheese of Quaker origin – cream cheese. Most cheese was for home consumption with commercial cheesemaking first emerging in upstate New York around Herkimer after 1857. Americans, however, were late to develop an appreciation of cheeses and no artisan tradition developed until the late 20th century. Factory production of cheese and butter essentially superseded the on-farm transformation of milk into cheese and butter. At the same time, the demand for market milk would increase, fueled by concerns touting milk's healthfulness. Fortunately a group of scientific advances made it possible to satisfy the new ever growing market. Among these was the Gail Bordon invention of the vacuum condenser in 1856, which made canned milk possible, Louis Pasteur's discovery that harmful bacteria could be destroyed by heating milk to 140 degrees between 1860 and 1865, the Frenchman DeLaval's 1879 invention of the first continuous milk separator, and an accurate and inexpensive technique for determining milk's butterfat content by Wisconsin's S. B. Babcock in 1892.

Historically, most farmers would milk their cows only in spring and summer when pasture grasses were lush. The introduction of the silo around 1875, where silage could be stored for feed year round led to year-round milking. Next came increasing attention to herd improvement, which became

Cows and the farm wife had a special relationship. This turn-of-the-20th century real photo postcard was sent to a cousin after mother's death "... she milked her for three years and thought so much of her." Now only Anna can milk "Spotty." *Collection: Mr. and Mrs. Michael B. Emery.*

Milk

We get milk from cows.
Cows live on farms
in the country.
We all drink milk.

19

On modern farms, by 1945 milking was done under controlled circumstances. Most milking was done by machine, but the cows would be "finished" by hand. Note the milking stool and the milk pail. The less formal boy with the milk pail was photographed in 1939. An early 20th century milker gives a cat a direct drink of milk. Chickens and ducks forage around the farmyard and especially the manure pile. *1945, The Cottage Collection, others, Collection: Mr. and Mrs. Michael B. Emery.*

general only after 1875. Over the years, bovine diseases were increasingly understood and controlled, especially tuberculosis.

As dairy farming became more and more demanding, it became increasingly male-dominated because women, in addition to their other duties, had often done the milking as well. Milking machines were introduced about 1875 but were crude and inefficient. It wasn't until the age of electricity that they were highly developed and in common usage.

There is a very interesting correlation between dairy production and pig raising: in areas where a lot of cheese and butter was produced pig farming was especially prevalent because the pigs were slopped with the whey and other liquid by-products of dairy or "creamery" production. A rule of thumb was that one hog could be fed for every four cows a farmer milked.

As with cattle, as the 19th century went on a concern grew with the quality of pigs raised. By mid-century the razorback hog, commonly seen on farms, were replaced by English breeds, including the Berkshire, the Suffolk, and the Duchess. After 1845, an American breed, the Chester White, developed in southeastern Pennsylvania, became popular.

Today, since most milk is sold as liquid, few pigs are milk fed and most are fattened on grain. While much of commercial pig farming has moved to the Midwest, pig farming and large scale fattening operations "feed lots" are profitable in the corn raising areas of Pennsylvania and other eastern states. Pigs are a perennially favored crop animal not only because of the versatility of the meat – used fresh and processed – but also because the meat can be produced at a much lower cost than beef. One pound of pork requires only half the grain that is need for the same amount of beef. Traditionally pigs were slaughtered on almost every farm, and butchering equipment survives in large amounts, as do the implements associated with most animal husbandry.

The eternal question for the dairy farmer is posed in the *Chester County Directory*, 1914; "Say, Mr. Farmer, Are You Making Money? Pure bred stock was the solution. *Collection: Mr. and Mrs. Michael B. Emery.*

While pig skins, like cattle hides, have commercial value, sheep are unique in that their chief value was for the product of their skin – wool, the fleece itself – rather than for mutton, their meat which was considered to be inferior. "For many years there was a strong prejudice against mutton, which softened and deteriorated rapidly, even when smoked or salted, hence the derisive term 'mutton-headed'". American sheep were seldom milked.

Like pigs and cattle, colonial sheep were usually expected to fend for themselves, but they were not very able to defend themselves and were subject to attacks by wolves and wild dogs. Early sheep were sorry-looking animals, small and slow to mature and their clip was two or three pounds of coarse single staple wool usually made into homespun for use on the farm. Writing to George Washington, Richard Peters, a gentleman farmer from Pennsylvania, noted "For some years hence this will not be a great sheep country." Why? "The dryness of our seasons burns up the pastures ... Our long winters are inimical to sheep ... We can have no succulent or green forage ... turnips are out of the question, [as] our snows and severe winters destroy or cover them."

After 1800 attempts were begun to introduce better sheep stock into America with the Leicester and Bakewell becoming popular. The greatest boost toward having purebred sheep came with the "Merino Craze" which peaked in 1816. Wool of ordinary sheep sold at 40 cents a pound while Merino wool sold for $2.00 or $3.00 a pound. Sheep brought as much as $1000 each. When the inevitable bust came, the animals could be bought for several dollars a head.

With the opening of the railroads, sheep raising, like beef raising, moved west although it remained common for individual farms to raise small flocks and, for a while, raising lambs for market became economically viable.

Each of these animal groups had specialized implements or tools associated with their culture and many of these, like oxen yokes, are collected today for their sentimental or aesthetic qualities.

LOOKING EAST FROM MACHINERY HALL, BETHANY ORPHANS' HOME WOMELSDORF, PA.

PURE BRED AYRSHIRE AND GRADE CALVES, MASONIC HOMES, ELIZABETHTOWN, LANCASTER COUNTY, PA.

Round Barn, Hershey, Pa.

Chester County farmer, Charles Plank, and his son pose in their farmyard *circa* 1944. The barn in the background had been built by the boy's great-grandfather in 1897. More substantially fenced dairy cows are seen on the farm operated by the Bethany Orphans' Home in Womelsdorf, *circa* 1910. Modern tile-sided silos dominate the Masonic Homes dairy operation near Elizabethtown, *circa* 1930; cows producing milk for Hershey chocolate bars graze near the dairy's famed Round Barn.
Collection Mr. and Mrs. Michael B. Emery.

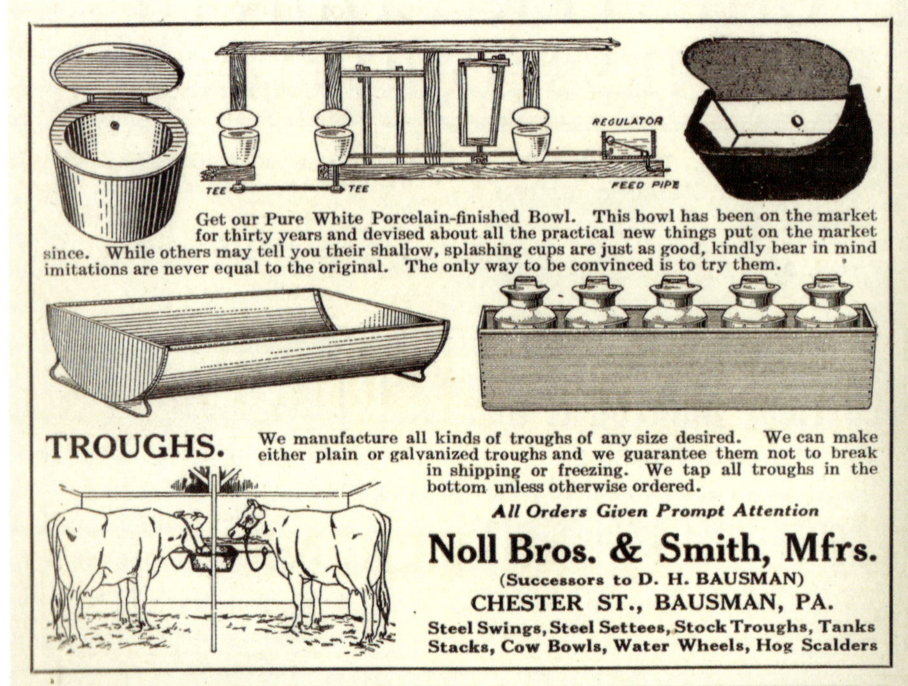

Modern farming meant factory-manufactured dairying equipment, as advertised by "Noll Bros. & Smith, Mfrs" in 1914, as well as veterinary medicines and specialty feeds. *Collection: Mr. and Mrs. Michael B. Emery.*

Fig. 4.—SECTION OF ICE-HOUSE AND DAIRY.

The ideal small farm dairy was located near ice-storage. Note the cylindrical butter churn, with its crank handle.

Early in the 20th century milk was delivered by truck. The Landis Valley Dairy promised "strictly FRESH MILK" that was "BEST for BABIES."

In a mid-19th century illustration, a farmer plows with a team of oxen while two men erect a fence. Specialized tools to fashion fence posts were common farm equipment.

Oxen, trained cattle of both genders, sure-footed and, well, "Strong as an ox," were often favored to pull both carts and wagons as well as to plow. The empty hay wagon drawn by two oxen is similar to one illustrated in chapter 1. The four-oxen team would have pulled a heavier vehicle and its contents. *The Cottage Collection.*

While cattle are used for both dairying and traction, most are raised for their meat. By the end of the 19th century, the largest commercial stockyards were located in Chicago and western beef came to dominate the American market. *The Cottage Collection.*

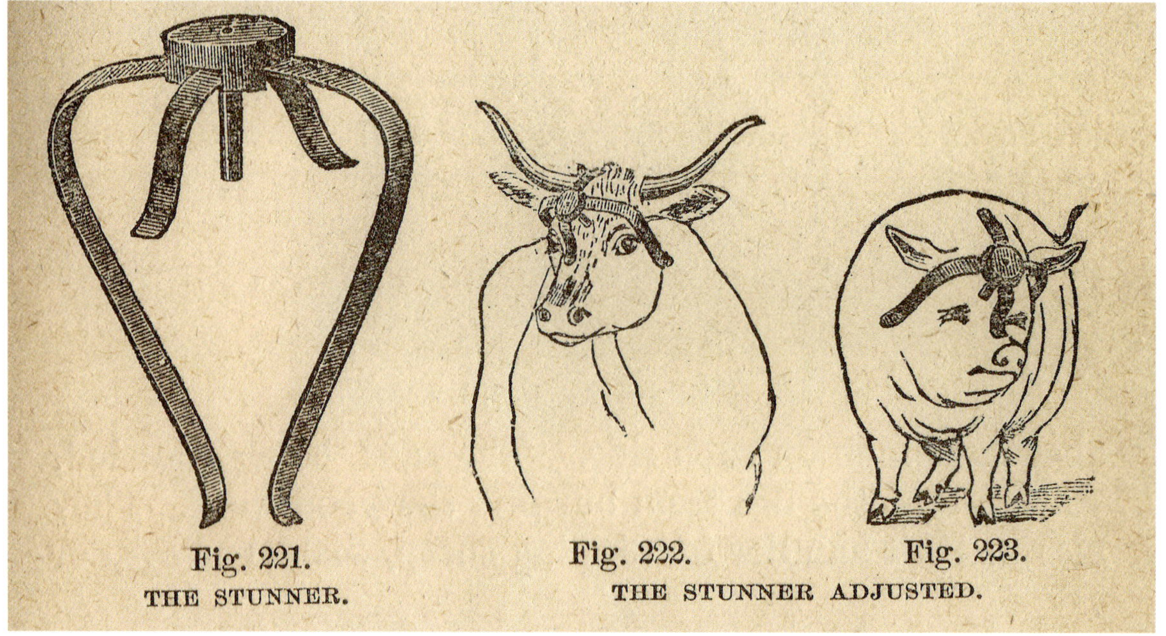

On many farms slaughtering cattle and hogs for meat was commonplace and all manner of butchering devices were in use. "With a view to avoiding all mishaps, saving pain, and leaving the operator free to sever the veins without embarrassment from the squealing and struggling victim, the design called "The Stunner," figure 221 has been invented. It fits over the head of the intended victim, as seen in figures 222 and 223, and a sharp blow on the plate over the forehead drives the pin into the brain, causing insensibility instantly, and death will not be long delayed. The use of such a mask is made compulsory in many countries of Europe. Such a contrivance is not only convenient, but humane, and appeals to the better nature of every man who is under the necessity of killing a dumb beast. As soon as the animal is struck, the throat is cut to insure free bleeding." (1887)

Raising hogs was common on most farms up through World War II. Many pigs were fed, not only with corn, but with slop, organic by-products. Hogs that became very fat and large were favored in the 19th century. "Stonewall Jackson" was pictured in 1878. Piglets and hogs are feeding in the barnyard. The buckets held by the farmer and his wife possibly held slop. *"Don't forget," The Cottage Collection. Piglets and hogs, Collection: Mr. and Mrs. Michael B. Emery.*

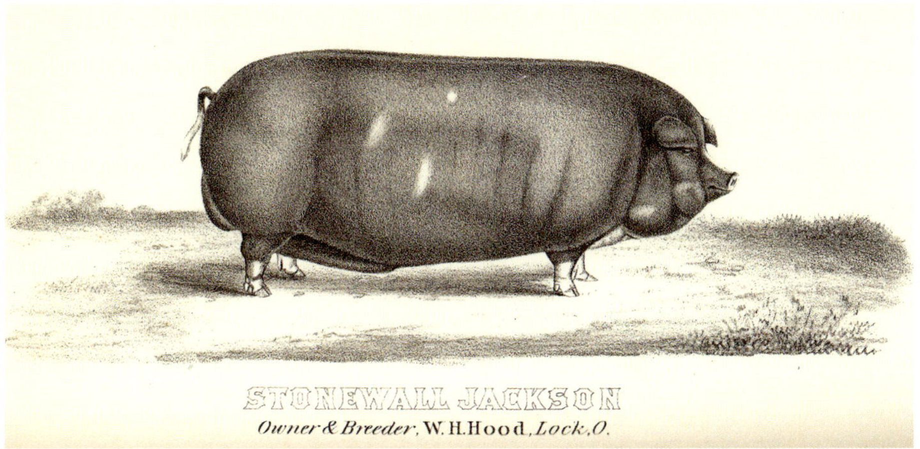

Hog butchering was common on farms – usually in November and December when the weather turned cold. One step in the process was to scrape the bristles off the hog before skinning. Hog bristles were prized for brush-making and used in plastering.

On many farms sheep were allowed to graze in pastures and in harvested meadows. These sheep in full fleece pose near a storm-felled tree on the Emery farm. Historically trees were allowed to remain in pastures as shade for animals. *Collection: Mr. and Mrs. Michael B. Emery.*

Dipping Sheep, Eastern Oregon.

Sheep are dipped before shearing to kill insects living in their fleece. The pastoral view, printed in Germany, dates from ***circa*** **1910.** *The Cottage Collection.*

Fig. 13.—FEED-BOX FOR SHEEP.

Why a feed box for sheep? "It is often inconvenient to go among the sheep in feeding them, and there is always trouble from scattering hay or feed about the enclosure or from the animals getting out by the open doors or gates. Figure 13 shows how to feed from outside." (1887)

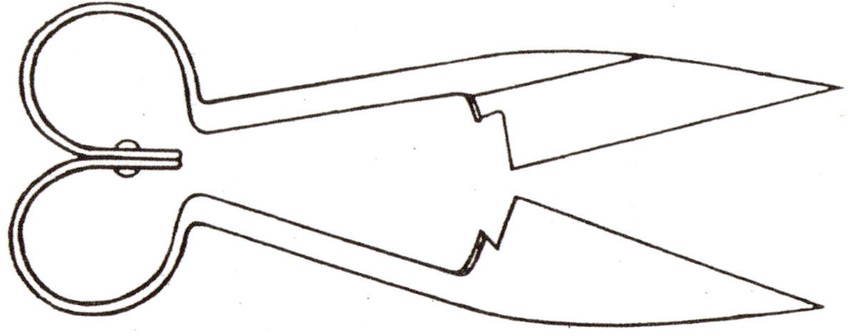

Sheep Shears

Sheep were traditionally shorn of their fleece with special shears designed to be used in one hand. In the mass-shearing scene, *circa* 1910, men employ these hand tools. In the post-World War II era mechanical shearing became very common. Traveling sheep shearers peddled their services to the owners of small sheep herds. *The Cottage Collection. Traveling shearer, Collection: Mr. and Mrs. Michael B. Emery.*

Many farms also had a few goats. Some were milked, but they were especially useful to consume noxious weeds. Goats will eat poison ivy to the roots. A fodder chopper and elevator stand in front of wooden silo, *circa* 1920. *Collection: Mr. and Mrs. Michael B. Emery.*

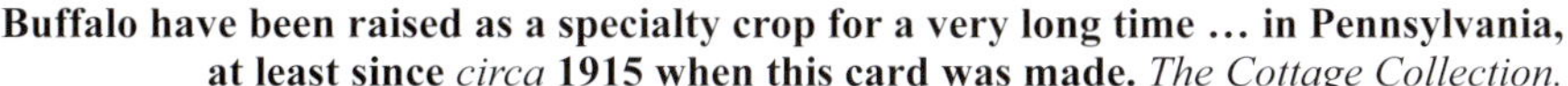

Buffalo have been raised as a specialty crop for a very long time … in Pennsylvania, at least since *circa* 1915 when this card was made. *The Cottage Collection.*

From the Landis Valley Collection

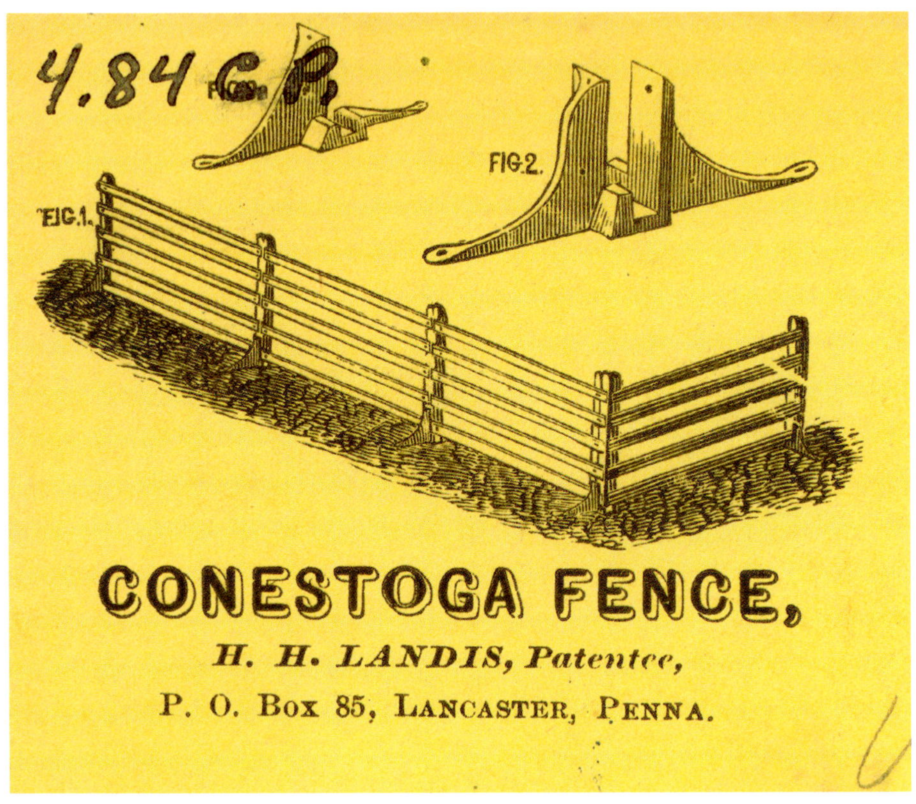

Flyer for a Fencing System. Henry H. Landis, the father of the founders of the Landis Valley Museum, did many things – always looking to strike it rich.

Augers and Mortise Axes. The tools were used together to create the slots in fence posts.

Mallets and Wedges. Used to split rails for fences.

Fence Post Cutting Machine.

Fence Hole Tamper. Used to pack earth around newly installed fence posts.

Post Hole Spade Heads.

American Line Back Cattle at Landis Valley.

Cow Bells, 18th and 19th centuries.

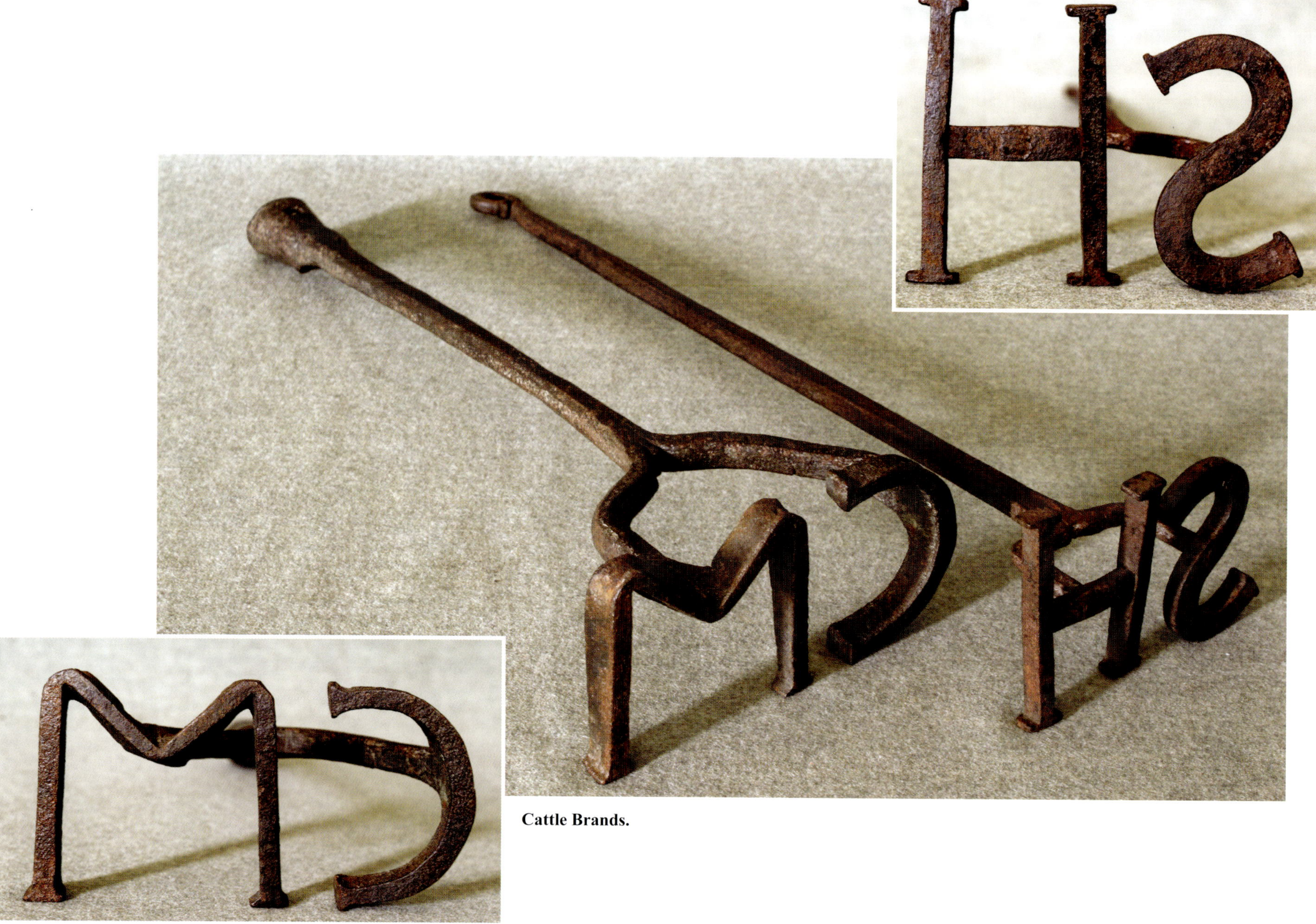

Cattle Brands.

Pitch Fork. Used to spread cattle bedding and to feed hay.

Feed Choppers. Used to chop pumpkins, turnips or potatoes before being fed to cattle.

Hydraulic Ram. Used to bring water into barns.

Light Plant. Generator used to bring light into a barn.

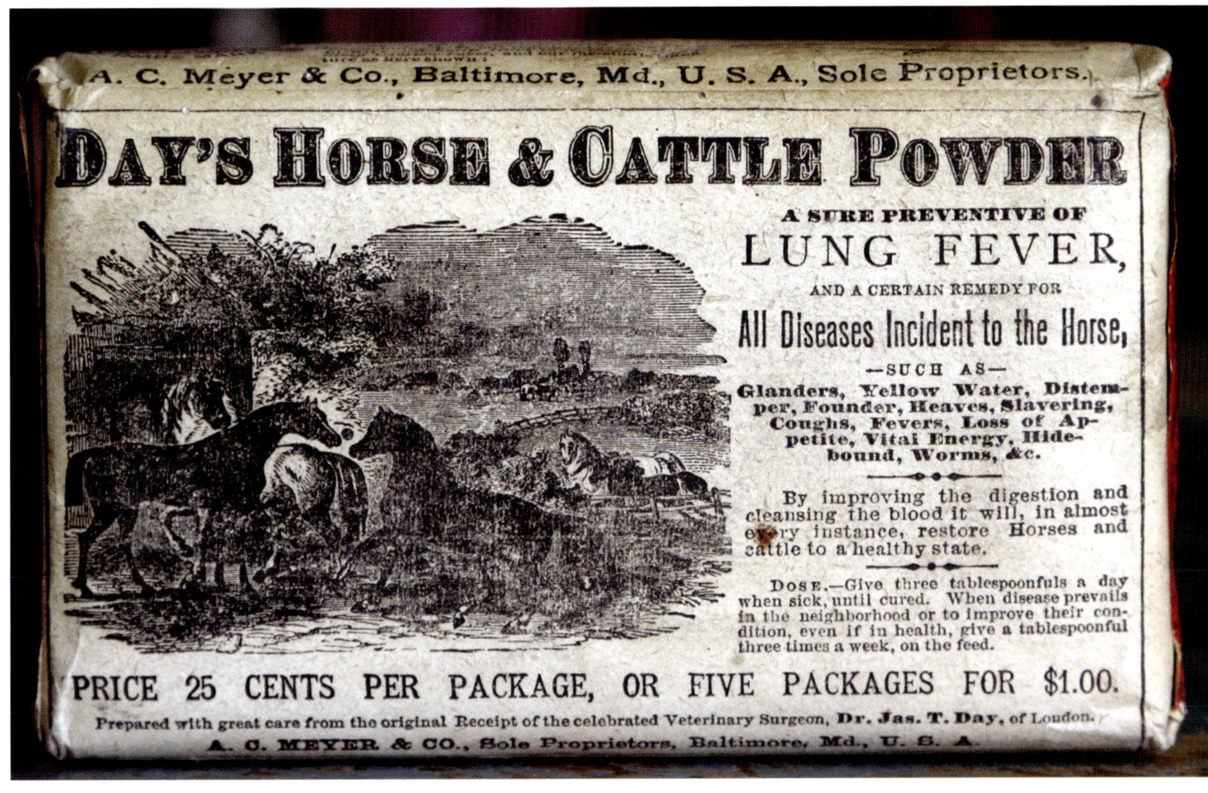

Veterinary Medicines and kindred chemical products, including margarine, the dairy farmer's enemy.

Cattle Hobble. Placed on the front legs, it kept an animal from wandering.

TO PREVENT COWS KICKING.

Cow Kick. The milker's best friend.

Milking Pails.

Milking Pails.

Milking Stool.

Milk Pan. Used to allow cream to rise so it can be used to churn butter.

Landis Valley Dairy Milk Bottle.

Milking Machine.

Milk Pails. Used to carry milk home.

Barrel Butter Churn.

Dog Treadmill.

Fig. 2.—WHEEL DOG-POWER.

Butter Bowl and Paddle with Butter Worker. Used to separate any fluids out of churned batter.

IMPROVED BUTTER-WORKER.—FIG. 1.

Butter Crock.

Butter Prints and Presses.

Butter prints and press.

Butter Box used to carry butter to market.

Cheese Molds.

Ruby and Ladd with Oxen Driver, Bob Siever, at Landis Valley.

Oxen Shoes.

An Heirloom Hog at Landis Valley.

Pig Sign. Probably from a butcher shop.

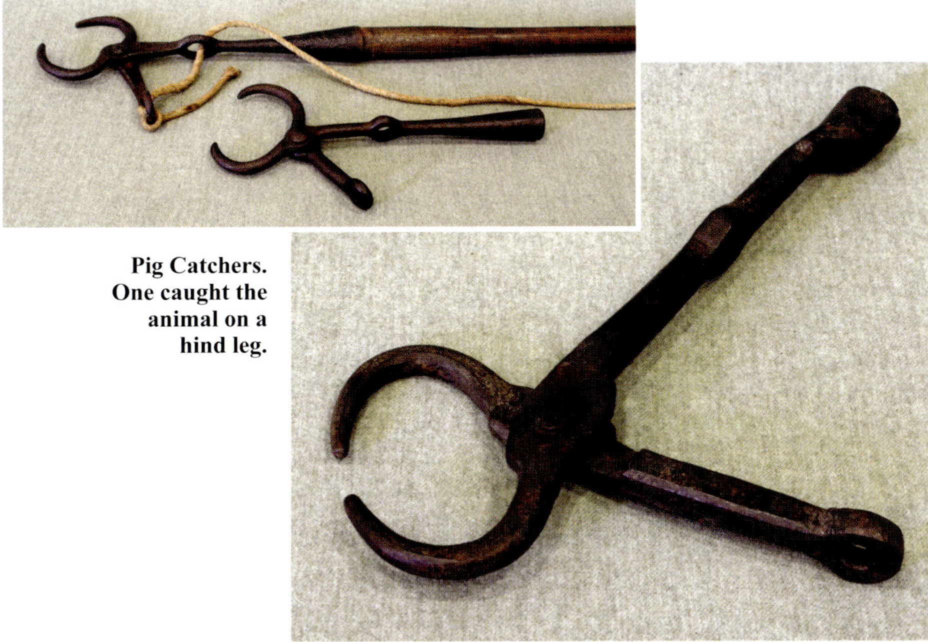

Pig Catchers. One caught the animal on a hind leg.

A Tunis Sheep at Landis Valley.

Sheep Shears.

Shepherd Crooks.

Model Manure Spreader.

Manure Hooks and photographer Craig Benner.

Manuring with Oxen.

Manure Spreader at Landis Valley.

Chapter 8

Horses and Mules

Before the age of steam, farm motive power was animal power. While small farms might use dogs to power treadmills to churn butter, the major draft animals were oxen, horses and mules. Oxen, bovines trained as draft animals, are also mentioned in Chapter 7. Mules were always work animals. Horses were, and remain, animals of work, recreation, and pride.

Prior to the late 18th century horses were rare on farms as oxen were more commonly used for farm chores and, until about 1860, controversy raged about the horses' role. Oxen were cheaper to acquire and less expensive to maintain. Oxen were also more sure-footed than horses on hillside farms.

Horses were mainly used for riding. Aside from walking, they were the only available means of overland transport. Owning a saddle horse was a luxury that most early farmers couldn't afford. From the earliest settlements, however, the wealthy took great pride in their riding horses. William Penn, for example, brought fine horses with him on both of his extended visits to his colony. On his second trip in 1699 he arrived with a much admired colt, "Tamerlane," sired by "Godolphin Barb," to whom many of England's best contemporary horses can trace their pedigree. Horses were not only highly desired, they were extremely mobile with the result that horse robbery was very common and the miscreant, if found, was harshly punished.

As roads became better during the first half of the 19th century, horse ownership became increasingly common on farms. Horses were quicker on the road than oxen and could be used to take produce to market. Draft horses are generally stouter and more heavily muscular than riding horses. The first great horse of American origin was a heavy animal developed by the Pennsylvania Germans in Lancaster's Conestoga Valley and used primarily to pull large Conestoga freight wagons, lyrically termed "the ships of inland commerce." The breed's ancestry is unknown but it is conjectured that it is a mixture of Flemish and English stock. In any event it was remarkably well suited to its work as described by a contemporary:

> The Conestoga horse had no growth of long hair or feathers between the knee and fetlock, as this would have been an unending source of trouble to the driver on muddy roads; a long tail also was a nuisance. The feet were moderately large but not flat, the top of the neck or crest well arched, body and legs rather short, temperament docile rather than nervous, movement forward stead and not wobbling, height to withers sixteen to

THE PRINCE OF QUALITY

Black Percheon Stallion.

This beautiful pure bred Stallion will stand for service at the stable of the owner on the Newport road between Hess mil and Intercourse, during the season of 1908.

Terms: $10 to Insure a Live Foal.

For further particulars and pedigree apply to the undersigned.

John Sigle, Owner

P. O. Gordonville, Pa.

Horses have always held a special place in the hearts of farmers. A celebrated award-winning Percheron, "Louis Philippe," grandiloquently named after a king of France, is celebrated in an agricultural magazine (1889) and you could rent the services of a "Prince of Quality" for $10.

seventeen hands and weight 1800 pounds or over in normal condition. The best colors in order were bay, black, gray, brown, chestnut, sorrel, roan, with not too much marking.

Eventually the Conestoga was replaced by the Percheron, a breed still known today.

Most farm work horses were of mixed parentage and their numbers grew enormously over the years as farm equipment became more complex and heavier. Reapers and other farm machinery needed traction and increasingly horses and mules were the answer. In 1910, for example, in Pennsylvania horses were the most valuable livestock appraised at $68,055,000 in comparison with cattle at $47,229,000.

The mule, a hybrid of the donkey and the horse, was popularized in America by George Washington who was gifted with one jack (male) and two jennies by the Marquis de Lafayette and the King of Spain in 1785. They soon became the preferred draft animal on southern farms and slowly found their way north where they were prized for field work. Smaller than horses, they are less expensive to feed and were especially used in cultivation.

Within the realm of the farm animal, perhaps the most storied are horses bred for racing, first trotters and then thoroughbreds for saddle racing. Out of this type of breeding emerged "Justine Morgan" in 1793, the sire of the

An anonymous draft horse is proudly displayed by his equally anonymous Pennsylvania owner (*circa* 1915); and George Diller Landis (1867-1954) as a country squire is posing with horse Billy and two of his hunting dogs, while holding a feed bucket in his hand. *Man with horse, Collection: Mr. and Mrs. Michael B. Emery.*

Farmer William Dries, an early owner of author Mike Emery's Berks County, Pennsylvania, farmstead, proudly poses with his team (*circa* 1910) wearing "fly chasers" designed to keep flies off the animals. More elaborate contraptions (although less successful ones) were available to keep the animals faces and ears clear of insects. A similarly attired horse draws a buckboard near a corn field. *Collection: Mr. and Mrs. Michael B. Emery.*

Morgan Horse line, some of which had distinguished careers on the race courses but most of which became carriage horses or the most favored farm draft animals.

While the age of steam had a minor effect on the number of horses kept on farms, the gasoline engine certainly changed the calculus. Farmers slowly converted to the tractor and away from the horse, although it has been estimated that by 1940 about two thirds of our farms still had horses and/or mules. They were especially useful during World War II when gasoline was severely rationed.

After the War work animals radically disappeared except among the Amish and some Old Order Mennonites who, along with a few other religious groups, still practice horse- and mule-based agriculture.

While mules were generally only considered to be work animals, horses were often prized. Many farmers had similar pride in their team that a modern farmer might take in his new pickup truck. Accordingly, once the age of photography came along, so too did many photographs of farmers posing proudly with their animals. Mankind's love affair with the horse didn't end when modern farming and transportation rendered it an anachronism. Today there are many hundreds of thousands, or perhaps millions, of horses in America. A few are used by farmers, some by cowboys, but more by carriage enthusiasts and pleasure riders. The race horse industry is still viable and now thousands of farms in America are devoted to the horse in all of its historic and modern applications.

Strong, well-groomed horses were prized for plowing and cultivating. Dunfermline is in Scotland, but the image was commonplace internationally. *Collection: Mr. and Mrs. Michael B. Emery.*

A single horse draws a farm cart driven by a costumed interpreter at the Landis Valley Museum.

Fig. 187.—A MANURE WAGON BOX.

Proud horses doing a less celebrated chore, pulling a "Manure Wagon Box." A manure hook lies in the foreground. "When hauling manure it is usual to drop it in heaps and leave it to be spread by a man who follows soon after. There are several methods of dumping the manure, but the most satisfactory is to use a manure hook. …The bottom of the sled or wagon should be formed of loose planks, each with its end shaved down to form handles. The side and end pieces of the box, though closely fitting, are not fastened together, so that they can be removed one at a time. One side or an end board is first taken out, and with a manure hook a sufficient amount of the load removed for the first heap." (1887).

Horses on treadmills provided power for a variety of early power-driven pieces of farm equipment. Here a team powers a thresher.

NEW YORK STATE AGRICULTURAL WORKS,

PATENTEES AND MANUFACTURERS OF

Railway Chain and Lever

HORSE POWERS,

Combined Threshers and Winnowers, Overshot Threshers,
Clover Hullers, Feed Cutters, Saw-Mills, Horse Rakes,
Horse Pitchforks, Shingle Machines, &c., &c.

ALBANY, N. Y.

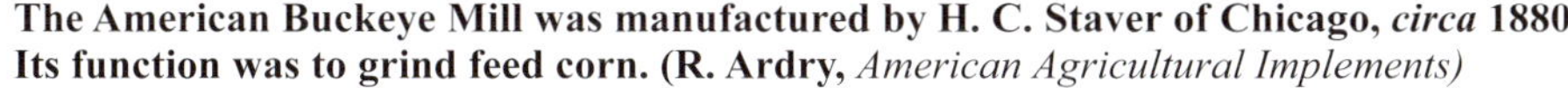
The American Buckeye Mill was manufactured by H. C. Staver of Chicago, *circa* 1880. Its function was to grind feed corn. (R. Ardry, *American Agricultural Implements)*

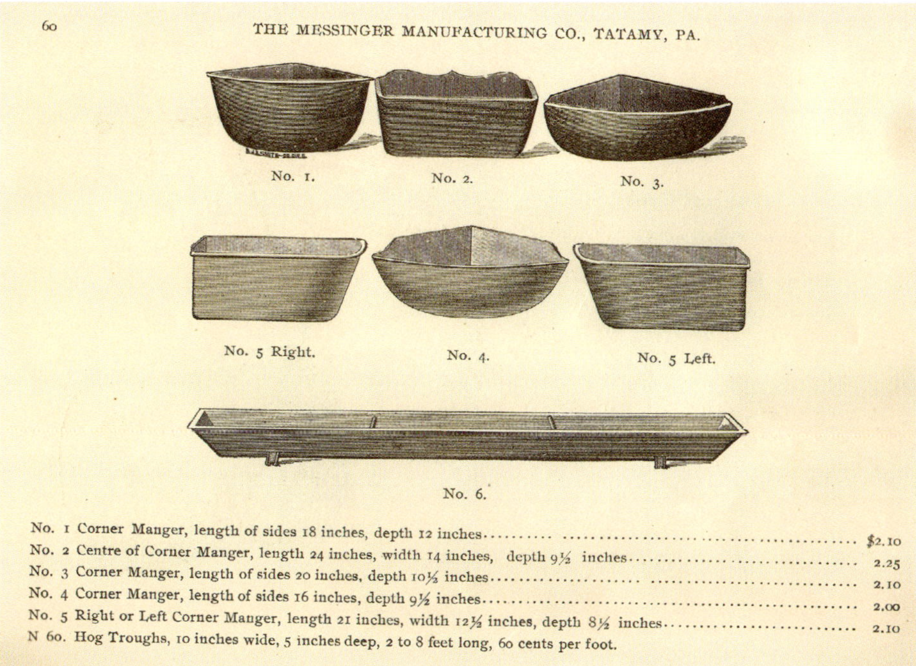
60 THE MESSINGER MANUFACTURING CO., TATAMY, PA.

No. 1. No. 2. No. 3.

No. 5 Right. No. 4. No. 5 Left.

No. 6.

No. 1 Corner Manger, length of sides 18 inches, depth 12 inches........ $2.10
No. 2 Centre of Corner Manger, length 24 inches, width 14 inches, depth 9½ inches........ 2.25
No. 3 Corner Manger, length of sides 20 inches, depth 10½ inches........ 2.10
No. 4 Corner Manger, length of sides 16 inches, depth 9½ inches........ 2.00
No. 5 Right or Left Corner Manger, length 21 inches, width 12½ inches, depth 8½ inches........ 2.10
N 60. Hog Troughs, 10 inches wide, 5 inches deep, 2 to 8 feet long, 60 cents per foot.

Farm equipment catalogs were popular after the Civil War and offered a wide range of products. Four mangers or feed boxes and a hog trough are featured in an undated catalog *circa* 1880.

Harness.
Saddles,
Whips,
Bridles,
Etc., Etc.
....
Repairing promptly attended to.

Honey Brook, Pa,......................189 .

M......................................

BOUGHT OF A. S. Troub & Son,
HARNESS MAKERS,
MAIN STREET, HONEY BROOK.

Harness makers were very important to the farmer and supplied many animal-related implements. Honey Brook is in Chester County, Pennsylvania. *Collection: Mr. and Mrs. Michael B. Emery.*

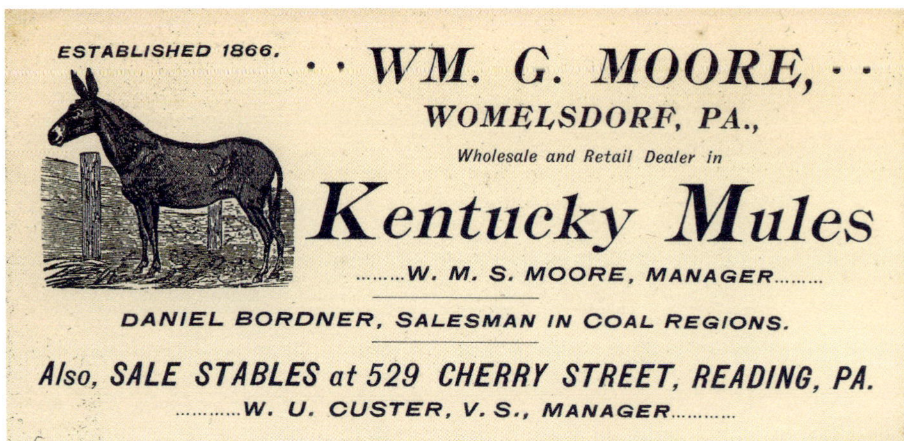

Mules were unsung draft animals of farms and mines. Many of the best, as with their more celebrated brethren, the race horses, were bred in Kentucky. Note that William G. Moore of Womelsdorf in the center of farm country had a salesman in the coal regions. *Collection Mr. and Mrs. Michael B. Emery.*

A three-mule team plows in the Lebanon Valley of Pennsylvania early in the 20th century. Six mules haul a load of agricultural lime in a photograph by H. Winslow Fegley at about the same time. A contemporary Amish farmer uses a five-mule team to harvest corn. *Plow scene, Collection: Mr. and Mrs. Michael B. Emery; Lime hauling, Schwenkfelder Library and Heritage Center; Amish, Courtesy Pennsylvania Dutch Convention and Visitors Bureau.*

Landis Valley Horses Wearing Full Bridles.

From the Landis Valley Collection

Haimes, Worn on top of the horse collar, it is attached to the bridle.

Horse Bridle.

Horse Collars and Haimes.

Horse Bits. Fitted into the horse's mouth, they were used in guiding the animal's direction and speed.

Running Bell. Used on horses when they were hitched to wagons, carriages, or sleighs.

Fly Chasers. Worn by horses to keep flies off.

Whips

Farrier's Box.

Farrier's Kit. Used in shoeing horses.

Buttresses or Hoof Trimmers (and details).

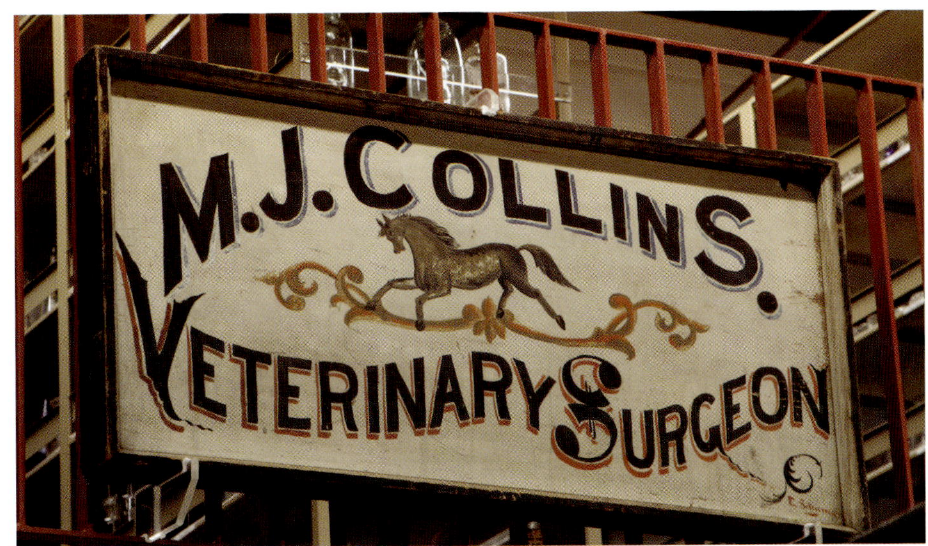

Veterinarian Sign.

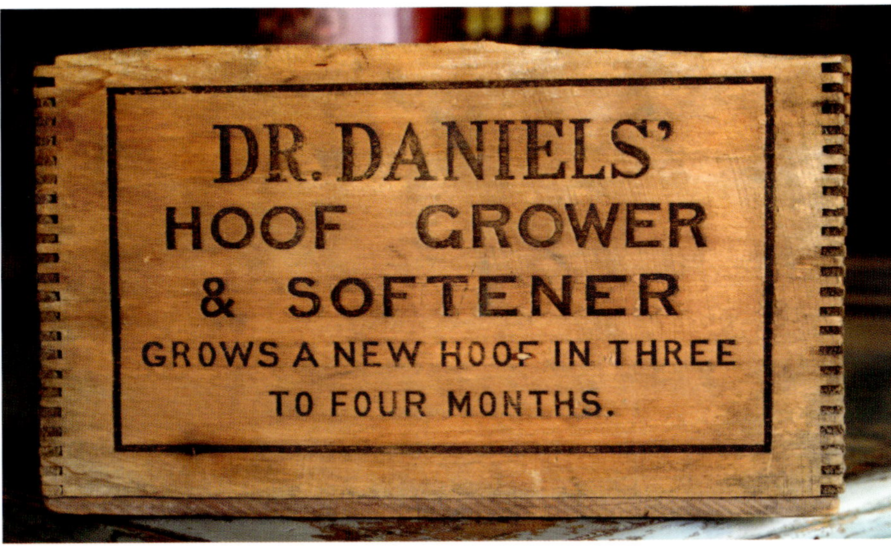

Country Store Sign and Product Being Advertised.

Horse Feed Packet Sample.

Model Horse Treadmill.

Horse Treadmill. Used in the pre-steam age to power many pieces of farm equipment.

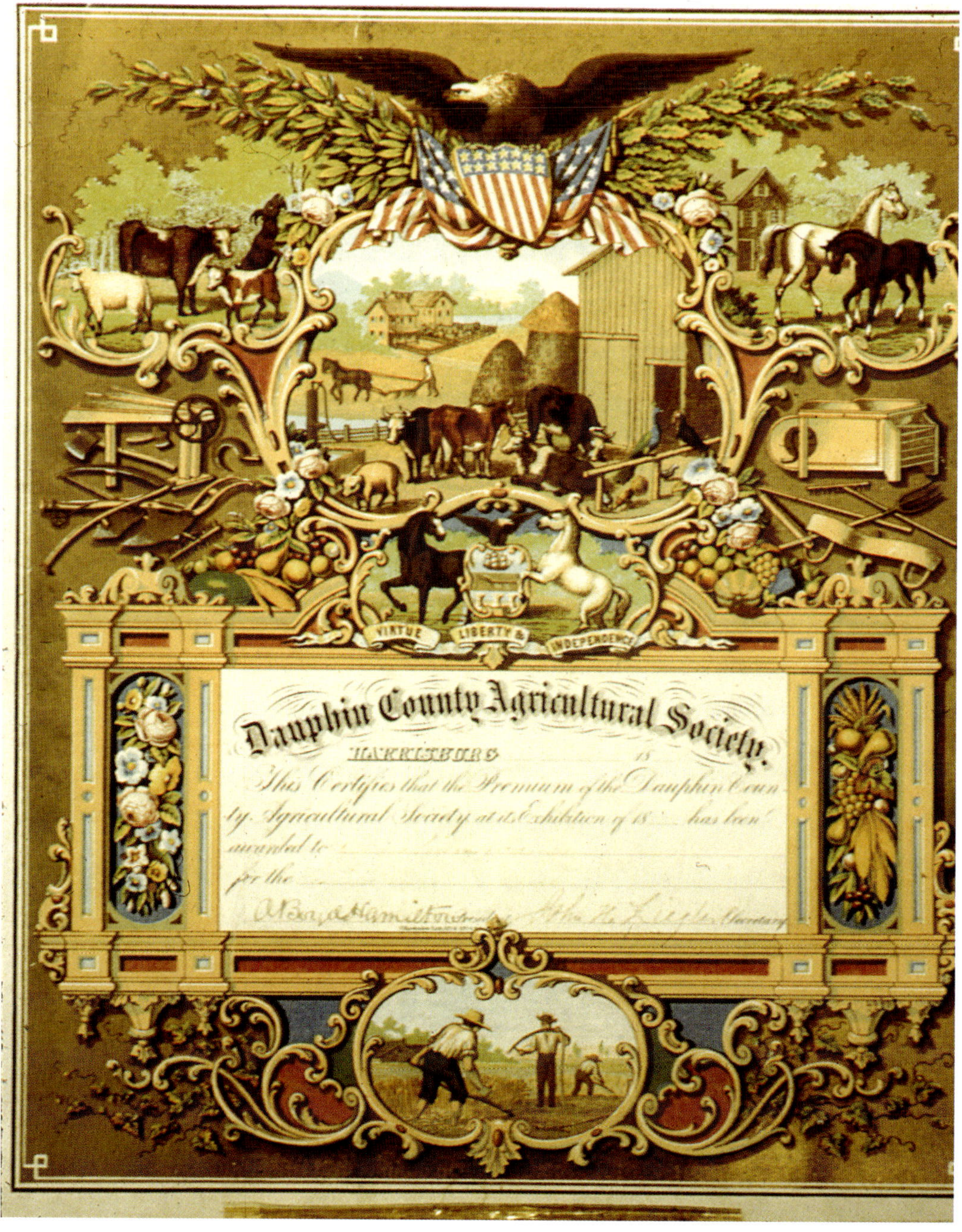

Membership Certificate, Dauphin County (Pennsylvania) Agricultural Society, *circa* 1860. A celebration of the importance of animals to agriculture. Note that horses are even featured on the Commonwealth's Coat of Arms (center).

Sources and Suggestions for Further Reading

Adams, Denise Wiles, *Restoring American Gardens: An Encyclopedia of Heirloom Ornamental Plants, 1640-1940.* Portland, Oregon and Cambridge, U.K.: Timber Press, 2004

Ashworth, Suzanne. *Seed to Seed: Seed Saving and Growing Techniques for Vegetable Gardeners.* 2nd ed. Decorah, Iowa: Seed Savers Exchange, 2002.

At the Greenmarket: You Say Tomato," *New York Magazine*. September 6, 2004, 127.

Beck, Berton E. "Grain Harvesting in the Nineteenth Century." *Pennsylvania Folklife* XXIII (Summer, 1974) 43-46.

Billard, Jules B. "The Revolution in American Agriculture." *National Geographic Magazine*. CXXVII (February, 1970) 147-185.

Boyer, Dennis. *Once Upon a Hex*: *A Spiritual Ecology of the Pennsylvania Germans*. Oregon, Wisconsin: Badger Books, Inc. 2004.

Decoteau, Randall. "Historic Harvest: Heritage Fruits and Vegetables." *New England Antiques Journal*. September, 2004, 51-54.

Ensminger, Robert F. *The Pennsylvania Barn: Its Origin, Evolution and Destruction in North America*. Baltimore: The Johns Hopkins University Press, 1992.

Fegley, H. Winslow. *Farming, Always Farming: A Photographic Essay of Rural Pennsylvania German Land and Life*.... Birdsboro, Pennsylvania: The Pennsylvania German Society, 1987.

Fletcher, Stevenson Whitcomb. *Pennsylvania Agriculture and Country Life, 1640-1940*, 2 vols. Harrisburg, Pennsylvania: Pennsylvania Historical and Museum Commission, 1950.

Gardner, Frank D. *Traditional American Farming Techniques*. Guilford, Connecticut: The Lyons Press, c. 2001.

Hedrick, Ulysses Prentiss. *A History of Horticulture in America to 1860.* New York: Oxford University Press, 1950.

Landis, Henry Harrison, "Diary," Manuscript Transcribed by Dawn L. Fetter, Landis Valley Museum Archives

Long, Amos, Jr. *The Pennsylvania German Family Farm.* Breinigsville, Pennsylvania: The Pennsylvania German Society, 1972.

Martin, George A., ed., *Farm Appliances. A Practical Manual*. New York: O. Judd Co., 1887.

Messinger Manufacturing Company. *General Catalog*. Tatamy, Pennsylvania: Messinger Manufacturing Comp, nd.

Partridge, Michael. *Farm Tools Through the Ages*. Boston: New York Graphic Society, 1973.

Richman, Irwin. *Pennsylvania German Farms, Gardens, and Seeds: Landis Valley in Four* Centuries. Atglen, Pennsylvania: Schiffer Publishing, Ltd., 2007.

Richman, Irwin. *The Landis Family: A Pennsylvania German Family Album*: Charleston, S.C., Arcadia Press, 2008.

Richman, Irwin. "The Pennsylvania-German Four Square Garden". *The Magazine Antiques*. July 2001, 92-97.

Sauers, W. Ray. "Fruit Harvesting and Preservation in Early Pennsylvania." *Pennsylvania Folklife* XXIII (Spring, 1974) 38-43.

Sears, Fred C. *Productive Orcharding: Modern Methods of Growing and Marketing Fruit*. Philadelphia: J. B. Lippincott Co., Copyright, 1914.

Stark, Les, *Hempstone Heritage I: In Accordance with their Wills: "All the Heckled Hemp She Can Spin"*, Morgantown, Pennsylvania: Hempstone Heritage, 2005.

State Board of Agriculture. Second Annual Report of the Pennsylvania Board of Agriculture For the Year 1878. Harrisburg, Pennsylvania: Commonwealth of Pennsylvania, 1878.

Steward, John F. *The Reaper: A History of the Efforts of Those Who Justly May Be Said to Have Made Bread Cheap*. New York: Greenberg, 1930.

Swift, Rodney B. *Who Invented the Reaper*. Chicago: Privatcly Printed, 1897.

Thompson, Ann Newlin, *et al. Germantown Green: a Living Legacy of Gardens Orchards, and Pleasure Grounds*. Philadelphia, Pennsylvania: The Wyck Association, The Germantown Historical Society, The Maxwell Mansion, 1982.

Weaver, William Woys. *Heirloom Vegetable Gardening: A Master Gardeners Guide to Planting, Seed Saving and Cultural History.* New York: Holt, 1997.

Wendel, C. H. *Encyclopedia of American Farm Implements & Antiques*. Iola, Wisconsin: Krause Publications, *c*. 1997.

Wendel, C. H. *150 Years of International* Harvester. Crestline/Motorbooks, 1981.

Index